Laser Light and X-Ray Scattering

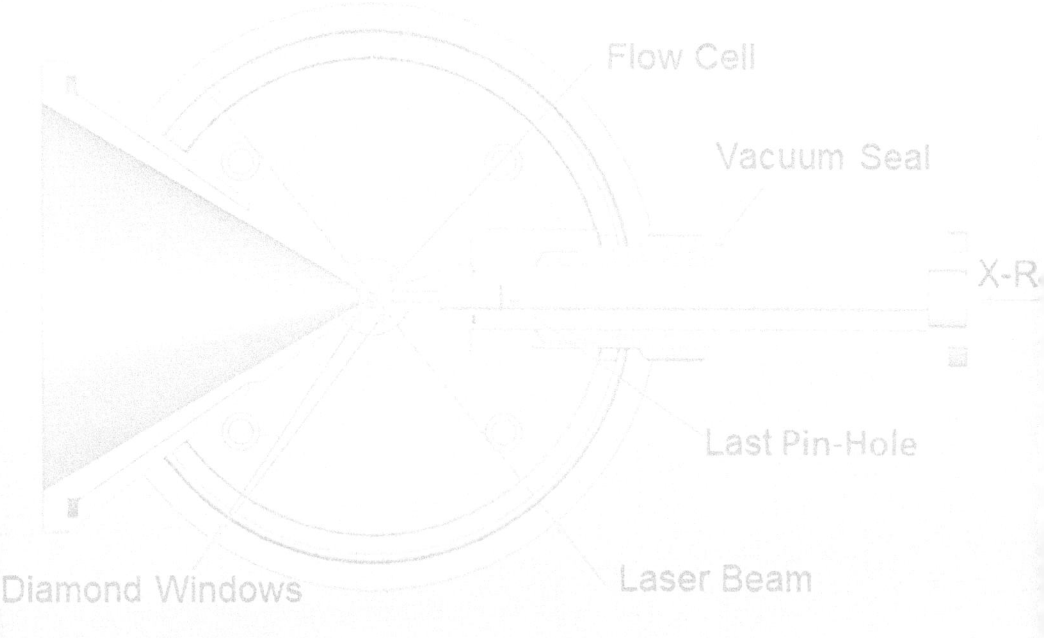

Laser Light and X-Ray Scattering

With Evolution of Electrospinning Technology in Biomedical Applications and Water Purification

Benjamin Chu

JENNY STANFORD
PUBLISHING

Published by

Jenny Stanford Publishing Pte. Ltd.
101 Thomson Road
#06-01, United Square
Singapore 307591

Email: editorial@jennystanford.com
Web: www.jennystanford.com

British Library Cataloguing-in-Publication Data
A catalogue record for this book is available from the British Library.

Laser Light and X-Ray Scattering: With Evolution of Electrospinning Technology in Biomedical Applications and Water Purification

ISBN 978-981-4968-88-1 (Hardcover)
ISBN 978-1-003-54074-8 (eBook)

Contents

Preface and Philosophy

Preface

My former students and I have written two review articles [1, 2]. One review deals with techniques of synchrotron X-ray scattering and laser cross-correlation function, while the second review is concerned mainly with the use of electro-spun nano-fibers for applications to water purification. The two reviews have a total of close to 500 references, bringing the two topics up to date on recent advances as related to those two topics. However, as the author for both reviews, meaning BC (or I), reads over those two reviews, the articles may be difficult for a beginning graduate student to appreciate the connections between the two seemingly different subject matters and to learn fundamental ideas as well as their applications. Thus, I now consider a teaching approach, trying to guide the interested graduate students as well as researchers who are less familiar with one or both of those two topics.

The development of the two review themes [1, 2] has their independent subject matter as they originate from different disciplines in physical/biological sciences and engineering. It also represents a current trend in cross-disciplinary training at institutions of higher learning. The two topics may represent an interesting example on interdisciplinary research and applications. In this respect, a short discussion on the perception of these two themes, their brief historical background and a reasonable justification on the future prospect of those two themes may be of interest. Clearly, we do not want researchers to enter a field of study which will become obsolete in the foreseeable future. I hope to convince the reader that the two review topics shall have good utilities for many years to come.

Philosophy

I completed my junior high school in Shanghai, senior high school (2 years) in Hong Kong, undergraduate (BSc, 2.5 years) and graduate (MSc & PhD, 3 years) studies in the United States. I was fortunate to have excellent research advisers in undergraduate, graduate, and postdoctoral studies. I was also very lucky to have met such wonderful teachers, *essentially by accident*. For most starting graduate students, who are looking for a good research adviser, it is often a challenge to actually find compatible research adviser(s) with mutual interest and benefits. Thus, I take the current opportunity (this introductory book) to add a few remarks in Chapter 1. Students often listen to other more senior students. However, it is also of interest to learn how to carry out his (or her) own independent evaluation in the selection of a good research adviser. Remember that the research team composition and personality of team members are often essential elements in his (or her) wish to receive a rewarding graduate education. Two final notes and a clarification:

- In our discussion on the topic in Chapter 2, we shall be concerned with the principles, dealing with synchrotron X-ray scattering, laser cross-correlation function, and the associated instrumentation techniques which can provide useful information for specific problems of interest. Some of those instruments are presented in Section 1.2.1 on *Synchrotron X-Rays*.
- In the development of more complex systems, it is likely that teamwork becomes essential. Thus, in addition to becoming an expert on our own specialization, it is important to be able to participate on the role as a member of the team, i.e., to have a good understanding on how the specific contributions can be integrated into the overall system. This is, for example, what Howard Hughes Medical Institute (HHMI) tried to accomplish years ago (Section 1.1.5). More importantly, we have noted the involvement of a broad range of disciplines even in those early time periods, including, for example, *law* and *humanities*. On a smaller scale, as we shall consider in Chapters 2 and 3, we need to integrate

specialized instruments into a synchrotron facility. Therefore, we must consider, for example, the specifications of the synchrotron beam lines, including space limitations and stability of the system, in order to make better use of those large facilities.

Gray scale versus color figures:

The book is grayscale in the printed form. However, most figures in computer presentations used a colored format. The descriptive sections have been unified.

References

1. Yimin Mao, Tianbo Liu, and Benjamin Chu. Scattering from Polymer Systems. In: Krzysztof Matyjaszewski, Yves Gnanou, Nikos Hadjichristidis, and Murugappan Muthukumar (eds), *Macromolecular Engineering: From Precise Synthesis to Macroscopic Materials and Applications*, 2nd edition, WILEY-VCH GmbH (2022). DOI: 10.1002/9783527815562.mme0022.
2. Xiaohan Xu, Yuqing Yang, Tianbo Liu, and Benjamin Chu, Cost-effective Polymer-Based Membranes for Drinking Water Purification, *Giant*, 10, 100099 (2022).

Chapter 1

General Background

1.1 Introduction

The reader is advised to first read the *Preface* and *Philosophy* pointing out the two recent reviews whereby two different topics on synchrotron X-ray scattering and laser cross-correlation function techniques (1) and the development of nanofibers, mainly produced by electro-spinning, are used for applications to water purification (2). To link the two seemingly different disciplines, I begin the discussion with an overall viewpoint whereby considerations based on even a broader perspective is presented, i.e., linking different traditional disciplines to aim at broader new objectives of benefits to society.

The book starts with a historical background and is partially based on a series of research presentations and a lecture course, centered on molecular aggregation and engineering, as given by Benjamin Chu (BC) at Stony Brook University. In the 21st century, having experienced the Coronavirus pandemic, followed by high-density population, and climate change, our economy is likely to be more disruptive but can also be remarkably dynamic. Together with a sense of unpredictable future, the capital-driven and technology-inspired Western economy, nevertheless, offers new opportunities. While the high-tech innovation on science and technology has emphasized on the digital world, with digital money and e-commerce, as well as gene-based bioinformatics, materials

Laser Light and X-Ray Scattering: With Evolution of Electrospinning Technology in Biomedical Applications and Water Purification
Benjamin Chu

ISBN 978-981-4968-88-1 (Hardcover), 978-1-003-54074-8 (eBook)
www.jennystanford.com

from macromolecular and supra-molecular aggregations, one of the broader themes of this book, can play a unique and indispensable role in the new environment.

The underlying general principles of the current approach involve a combination of basic and applied research with dual objectives on fundamental understanding (basic) of macromolecular materials that can be used to produce specific products (applied). The pathway takes into account of the university-industry-government relationship as an essential element in the process to achieve the dual goals. It should be noted, from the very beginning for the readers to consider the importance of including the university-industry-government pathway because university and industry are within our society, being governed by political institutions. Financial support, such as government grants, and implementation of regulatory rules, including environmental controls, shall invariably involve government agencies. Then, the more important themes of benefits to society, such as the purification of water and air, should involve the government, in addition to university and industry, often playing a key role in the process.

1.1.1 Preparations for Students to Meet New Challenges

The book tries to take into account of the needs of readers (aimed mostly for beginning graduate students) to have a broad perspective, in order to tackle new problems. It intends to match ability with interest and emphasizes on emerging fields. Aside from overlapping themes, the presentations in the book are associated primarily with those disciplines, as denoted in **bold characters below**, being connected to the two topics of current interest. A sketch on *interdisciplinary topics* as related to our current interests is as follows.

- Cellular and molecular biology
- **Chemistry – synthesis, analytical & physical**
- **Chemical and biomedical engineering**
- Computer science
- **Materials science & engineering**
- Electrical engineering
- Biological & condensed matter physics

The current chapter (Chapter 1) provides an overall description of the topics to be covered. The topics are being shown in bold characters in the *interdisciplinary fields* as being presented above. Chapter 1 includes an overview of the book and the historical perspective with respect to the subject matter. It is also of interest to know that many materials of years ago, especially those related to fundamental scientific knowledge, can actually be used to form an essential component of the subject matter for future studies. Thus, hopefully, the reader can benefit from this book, in combination with the two review articles, including references therein, and then proceed to develop new ideas for future research and technology development. From a slightly different perspective, the author (BC) hopes to introduce his research experience as a complimentary issue to (student) readers. It does not deal with the technical content but expands toward a more philosophical outlook on the manner in which we deal with in our university studies, as shown in Section 1.1.2, "Notes on Research Adviser."

Starting with Section 1.1.2, "Notes on Research Adviser", it is important to point out, early in the presentation of this book, that there is a relationship between the changes in our society, including our universities, the disciplines of interest, the educational approach in our schools and universities. Thus, Chapter 1 encompasses those changes, in addition to the technical presentation of later chapters. The philosophy, together with my personal experience as a postdoctoral student, and the technical presentations in later chapters, are different topics. Nevertheless, they are being linked together because the manner with which we approach our research is affected by the changes in our society.

1.1.2 Notes on Research Adviser

In selecting a research adviser, the author (BC) often thinks of his own experience with his postdoctoral research adviser, the late Professor Peter JW Debye at Cornell University. On several occasions in the beginning of his postdoctoral tenure, BC was first surprised to find how Professor Debye, primarily a theorist, knew so much of the experimental details. However, he also sensed a lack of enthusiasm in Professor Debye during those early discussion occasions. Later on, Professor Debye simply brushed BC off from his office, if the discussions dealt only with technical questions. BC changed his approach and learned a great deal from then on when the sessions started to deal, primarily, with discussions on problems as related to basic underlying principles or limits of applied technology as well as their impacts and future to science and society. On this issue, there is an article by Nobel Laureate Robert Lefkowitz who has shared 10 rules on the art of scholarly mentoring [1].

Finally, it is worthwhile to mention Warren Buffett's four words of wisdom: *"hire well, manage little,"* meaning that we should try to pick the right person for the job and then let the person develop his (or her) own program, based on a broad perspective. Buffett's wisdom is applicable for business, but it may also be suitable as a guideline for us in our own personal development, e.g., the student and the research adviser are looking for a compatible match. The important objective on a higher level does not simply deal with technical know-how but also the training for the student to be able to distill out the basic principles governing the specific topic. A student should not look for a research adviser whose primary interest deals with the search for a person to carry out routine technical work. Buffett has 4 more words of wisdom: "Do what you love" which we should always keep in mind.

In the current environment, research and development in university and industry, often being supported or guided by government agencies, do get integrated more closely with society, especially for topics of more immediate benefits to society. Thus, the outline of this book has five chapters:

Chapter 1, the *key* introductory part, provides an overview of the book on *functional soft matter of benefits to society.* It tries to consider the new economy as affected by COVID-19, high-density population, and climate change but also the *opportunities* in the new environment. Chapters 2 to 4 are used to fill out some of the details, while Chapter 5 aims for us to briefly look forward into the future.

Chapter 2 deals with the techniques of *laser light/synchrotron X-ray scattering*, leaving most of the specific example applications to Chapter 4.

Chapter 3 is concerned with *physical techniques on electro-spinning and electro-blowing technology*, as being used in *filtration membranes*, again leaving specific example applications to Chapter 4. It should be stated that the development of filtration membranes *for water (and air) purification* has a very broad scope. The use of electro-spinning and electro-blowing technology covers only one approach, i.e., dealing with the formation of nanofibers. There is a broad range of approaches dealing with how filtration membranes can be formulated for specific applications.

Chapter 4 provides example applications based on Chapters 2 and 3. More importantly, it deals with an overall discussion on the complementary nature of the technology in Chapters 2 and 3. In Chapter 4, it also tries to fill in more details on why *water (and air)* are and shall continue to be *key issues* of concern for the well-being of our society in the future.

Chapter 5 provides an optimistic and brief outlook toward the future, emphasizing the need and coupling between basic knowledge and practical applications.

1.1.3 More about Chapter 1

Chapter 1 provides a background on nano-science and -engineering, especially with respect to research opportunities as related to the applications of laser light and synchrotron X-ray scattering, which will be presented in Chapter 2.

While one of the general themes is on functional soft matter of benefits to society, our emphasis is on the topics related to macromolecular science and engineering. In particular, the applications of nano-fibrous membranes for water purification and a brief extra discussion on air filtration will be the examples for the applications. In the presentation below, it should be noted that some of the figures came primarily from lectures given by BC, with minor modifications. A limited amount of overlap from one chapter to another chapter is intentional in order to link different chapters together. It should also be noted that the last subsections in Chapter 1 are closely related to Chapter 2, in order to provide a smooth transition from general discussion to specific instrumentation development.

As has already been mentioned briefly, the laser light scattering and synchrotron X-ray small-angle X-ray scattering (SAXS)/ wide-angle X-ray diffraction (WAXD) instrumentation is intended for studying the structure and dynamics of materials. A brief presentation on the future in research on macromolecular aggregation and engineering, one of the more general themes of our current interest, is presented below.

1.1.4 Future in Research on Macromolecular Aggregation and Engineering

In considering the functional soft matter of benefits to society, the perception of the future in research and applications on the subject matter of macromolecules is of importance. Opportunities offered by combining the strength in *biology* and *chemistry* with the processing techniques in *engineering* and the characterization studies based on *physics* can help us achieve our goals and concerns of the current two topics (Chapters 2 and 3).

In the presence of COVID-19, high-density population, climate change and political turmoil, our economy tends to become somewhat disruptive, but can be remarkably dynamic and provide new opportunities, even though it may contain, invariably, unpredictable elements in a capital-driven and technology-inspired society. While the current popular trend in our society deals with the digital world and gene-based bioinformatics, as has been mentioned earlier, the two themes (Chapters 2 and 3) of our interest shall remain to be able to make important contributions to our society for many years into the future.

Let us consider some facts on resources in general. We think of the

- **衣:** *clothing* we wear - textile industry (e.g., smart clothing),
- **食:** *food* we eat - agriculture (e.g., genetic engineering),
- **住:** *house* we live in (e.g., optimal use of space and materials), and
- **行:** *ways we move and communicate* – (e.g., transportation and Internet communications).

More importantly, we need the *air* we breathe and the *water* we drink. Thus, water is and shall remain one of the essential and most important resources for the well-being of our lives in this world.

1.1.5 New Directions in Science/Engineering Education

Inter-/multi-disciplinary approaches in education were actually started years ago. Such needs came from more complex problems that required us to cut across traditional disciplines. Thus, research universities started to consider problem-oriented disciplines *without department*. The goals were in terms of *projects*. Examples involved cluster hiring of faculty with complementary expertise in different departments and then provided them with *common space*. Such practices started more than 20 years ago by private universities, such as Rockefeller University and Stanford University; state universities, such as the University of Wisconsin and the University of Michigan, and colleges, such as Haverford College.

In the late 1990s, a group of Stanford faculty, led by Steven Chu and James Spudich, organized a grass root effort to initiate a bold inter-disciplinary research and teaching program in the areas of bioengineering, biomedicine, and multidisciplinary macromolecular science & engineering. It also operated across the Schools of *Humanities* and Sciences, Engineering, Medicine, Earth Sciences, and the School of *Law*. The Bio-X Program and the concept of a radical new building, now known as The James H. Clark Center, started about 25 years ago. The vision of a multidisciplinary, collaborative community of science, consisting of biologists, chemists, computer scientists, engineers, and physicists who, with intellectual and financial independence, can work in small research groups, to nurture the forefront of science, later known as the Howard Hughes Medical Institute (HHMI) was established around 2006. Thus, we can see from one of the best examples that changes in format in higher education have taken much time and effort. In the meantime, the number of administrators in universities has increased substantially, partly because of increases in regulations imposed by numerous (often government) agencies. An increase in the number of administrators in a university is always a concern. However, this topic is outside of the current scope of our discussions.

The brief presentation above shows that with new challenges, the problem-oriented approach, involving multiple disciplines, is inevitable. At the same time, it is important to realize the need for fundamental understanding with practical applications. The team formation implies that we seldom work alone but require mutual understanding and respect with coworkers. At the same time, there are often some forms of bureaucracy, sometimes in terms of hierarchical bureaucratic rules, coming into our ways to deal with.

Aside from internal driving forces within the United States, it is also worthwhile to have some idea on the competition with the outside world. Let us use the car market as an example. Based on information from the August 7, 2006, issue of *Business Week*:

- In America, it was 750 cars per 1000 while the U.S. population was about 300 million.
- In China, it was 8 cars per 1000 with a population of 1.4 billion.
- In India, it was 6 cars per 1000 with a population of 1.1 billion.

In 2021, it was greater than 800 cars per 1000 for the USA and 200 cars per 1000 for China. Thus, China has about the same number of cars as the USA and has become a major car market. Under the current conditions, we can see that it is a tough challenge to nurture creativity which is a very important element in a competitive society.

1.1.6 Salient Features to Meet New Challenges

Based on the previous discussions (Sections 1.1.3–1.1.5), the noticeable features are summarized as follows:

1. To have a broad basic training to tackle new problems: the idea started years ago as has been briefly discussed in Sections 1.1.1–1.1.5.
2. It is better to match ability with interest. A personal interest in doing work that makes one feeling good should play a role in the preparation. Of course, one of the aims in education is for students to realize the opportunities available in the new world so that one can make more suitable and reasonable choices.

3. We should remember to train ourselves to nurture creativity, which is essential in a competitive world.

In Section 1.1.2, the important consideration is to be able to distill out the basic principles governing the specific topic. Section 1.1.5 considers the need which came from more complex problems that require us to cut across traditional disciplines. Thus, we realize that in considering the emerging scientific opportunities at interfaces among biology, chemistry, materials, and physics, *macromolecules and macromolecular aggregations*, which cover a range of interdisciplinary topics, as represented, primarily, by ***heavy italic*** symbols below, play an indispensable role in our well-being and are of benefits to society:

- Cellular and Molecular Biology
- ***Chemistry: Synthesis, Analytical, and Physical***
- ***Chemical and Biomedical Engineering***
- Computer Science
- ***Materials Science and Engineering***
- Electrical Engineering
- ***Biology and Soft Condensed Matter***

1.2 General Background for the Presentation Below

The presentation below provides an overview of the type of experiments that can be carried out by using laser light and synchrotron X-ray scattering with samples primarily related to those topics under current discussion. However, synchrotron X-ray techniques are useful over a *broader range of applications.*

The Advanced Polymers Beam Line at X27C of the National Synchrotron Light Source (NSLS), though no longer exists, is being used to illustrate a few key points in instrumentation development for synchrotron X-ray scattering. While NSLS has been replaced by NSLS-II, the basic design on many key components of such X-ray instrumentations remains essentially the same. Furthermore, beam lines often have associated instruments to control or/and measure other related physical variables of interest. Thus, the overview below tries to give a general background with additional details on instrumentation and applications that shall be presented in later chapters (Chapters 2 and 3).

It is noted that synchrotron X-rays are much more powerful than laboratory X-rays. Thus, it changes the way X-rays can be used as a "light source". Laser light and X-rays are forms of electromagnetic radiation. While lasers have coherence, synchrotron X-rays initially have mainly high intensity. However, new synchrotron facilities are capable of carrying out X-ray photon correlation, opening up new opportunities to study dynamics of small systems. In the illustration below, more emphasis has been given to demonstrate the importance of associated instrumentation which can be used in combination with synchrotron X-rays in order to expand the range of measurements that are being used to study macromolecular and supra-molecular systems. For synchrotron X-rays, due to its continuous spectrum, high flux, brightness, and coherence, spectroscopic, scattering, imaging, and dynamic studies become feasible. In the current presentation, our emphasis will be on scattering and dynamic studies.

1.2.1 Synchrotron X-Rays

Synchrotron X-rays have many very good characteristics, with a continuous spectrum, high flux, brightness, and high coherence. Synchrotron X-rays can be used to study both structure and dynamics of materials, including topics of current interests in macromolecules and macromolecular assemblies. Generally, we can consider that synchrotron X-rays can cover a range of information, including spectroscopy (energy), scattering (momentum), imaging (position) and dynamics (time).

For X-rays, the synchrotron characteristics can best be understood in terms of its intensity, as expressed by the following terms:

- **Brightness**, in units of photons/second mrad^{-2} mA^{-1} per unit bandwidth
- **Brilliance**, in units of photons/second mrad^{-2} mA^{-1} per unit bandwidth *per unit source area.*

Table 1.1 shows a comparison of X-ray sources versus project duration.

Table 1.1 X-ray sources versus project duration

Source	Brilliance	Experiment time	Project duration (as an example)
Sealed X-ray tube	10^8	10 h	10 months
Rotating anode tube	10^9	1 h	1 month
NSLS bending magnet	10^{13}	40 ms	about 4 min
Insertion device	10^{15}	0.4 ms	about 2.4 s
APS[a] (and newer beam lines)	(>)10^{19}	4 ns	0.24 ms

[a]APS, Advanced Photon Source, Argonne National Laboratory, Lemont, Illinois.

The computation in Table 1.1 is rather trivial. The reader can easily see from the difference between a sealed X-ray tube and a rotating anode tube on how the experiment time and the project duration can be computed. BC would like to see the reader (primarily the graduate students) to calculate the rest of the values, associated with a variety of synchrotrons. When we plan to carry out experiments in a synchrotron facility, we need to make such computations so that we can decide the experiment time needed for the proposed project. While it is wonderful to be able to take advantage of synchrotron light sources, possible radiation damage should always be kept in mind for the specific (especially biological) sample of interest.

We can realize that the change in brilliance from a sealed X-ray tube to synchrotron radiation has provided us with new opportunities in our experimental studies. As the X-ray radiation is a form of electromagnetic radiation, it is not only the increase in magnitude of the X-ray intensity, but also the coherence property associated with the X-ray radiation, even though synchrotron X-rays are not quite like lasers, in the normal format with which we are familiar.

With new synchrotron facilities, *X-ray photon correlation spectroscopy* (XPCS) has become a technique within our reach. At the same time, in order to take full advantage of such facilities, not to mention the availability of very-high-intensity X-ray sources for small-scale laboratories, it is also worthwhile to consider portable instrumentation that can be brought to the synchrotron facility to measure physical parameters of interest for research problems under investigation. For this reason, a few supplementary instruments are presented here to demonstrate that in order to make best use of national synchrotron facilities, the researcher may be interested to learn and to develop compatible portable instruments that can not only be incorporated with synchrotron beam lines, but can also be used at home institutions.

1.2.2 Supplementary Instrumentation for Synchrotron X-rays

Let us first start with a few instruments, such as a dual-chamber temperature-jump apparatus (Figure 1.1), a rheo-X-ray apparatus (Figure 1.2), and a simultaneous X-ray and Raman scattering instrument (Figure 1.3), as well as a high-pressure cell permitting combined synchrotron X-ray and laser light scattering measurements (Figure 1.4).

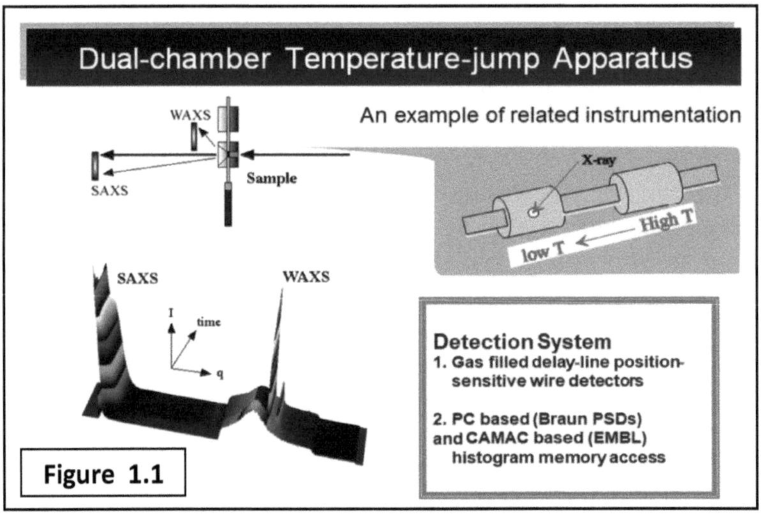

Figure 1.1 Dual-chamber temperature-jump apparatus. With high-flux X-rays, time-resolved studies are feasible. An example is to measure structural changes of materials after a sudden temperature change. The different shades for SAXS and WAXD profiles in the lower left section shows a 3 dimensional I-time-q plot, revealing how intensity (I), as a function of q, changes with time in a temperature jump experiment.

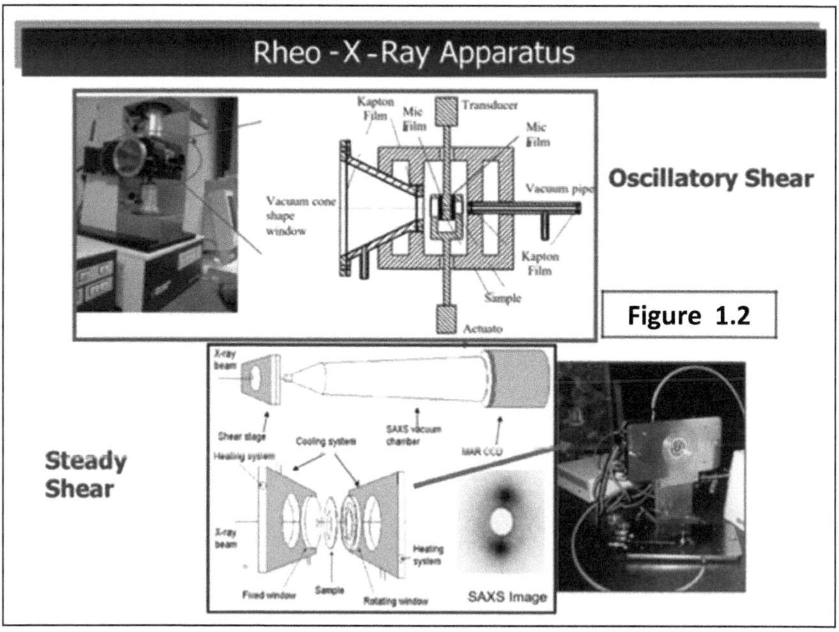

Figure 1.2 Rheo-X-ray apparatus. There are two formats, involving oscillatory shear and steady shear measurements. In addition to the schematic figure, there is also a corresponding picture showing how such an instrument looks like. As the apparatus needs to be inserted into the synchrotron X-ray beam path, between the synchrotron X-ray collimation system for the incident X-ray beam and the detection system to measure SAXS/WAXD X-rays from the sample during oscillatory or steady shear, instrumentation space allocation has to be considered in advance.

It should be noted that the first two supplementary instruments (Figures 1.1 and 1.2) are used to perturb the sample of interest and then let synchrotron X-rays to measure the corresponding changes. The third instrument (Figure 1.3) is to combine Raman spectroscopy (as another additional probe) with synchrotron X-ray measurements. The third instrument (Figure 1.3) is shown as a picture in Chapter 1. It has a line drawing, being presented in Chapter 2 under *Simultaneous Raman/Synchrotron X-ray Setup*, partially intended to encourage the reader to look for connections between Chapters 1 and 2.

Several units of such instruments were developed for use both at Stony Brook and also at synchrotron light sources, including the Advanced Photon Source (APS) in Illinois. A few units are described in Chapter 1.

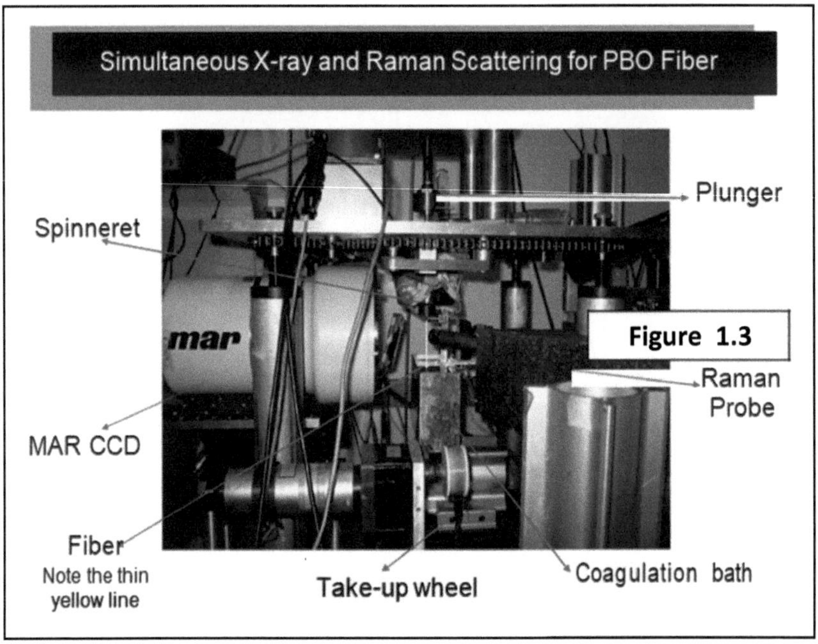

Figure 1.3 Simultaneous X-ray and Raman scattering for fiber spinning studies. The example phenylene benzobisoxazole (PBO) fiber, known as ZYLON by Toyobo in Japan, is the light-colored vertical line in grey-scale (and can be seen as a thin yellow vertical line in the colored presentation). The Raman probe is pointing at the fiber with the Raman spectrometer being located elsewhere. The MAR CCD is the X-ray area detector, looking at the fiber, passing through the spinning unit. Thus, such a setup is more complex because it involves a combination of not only two types of measurement units (X-ray scattering and Raman scattering), but also a fiber spinning unit which has a coagulation bath.

We start with a dual-chamber temperature-jump apparatus, as shown in Figure 1.1. The instrument is designed to carry out simultaneous SAXS and WAXD measurements when the sample of interest is undergoing a temperature jump. The temperature jump can be either up or down.

Note the top and side views of Figure 1.4, together with high pressure generation, control, and measurements of this instrument.

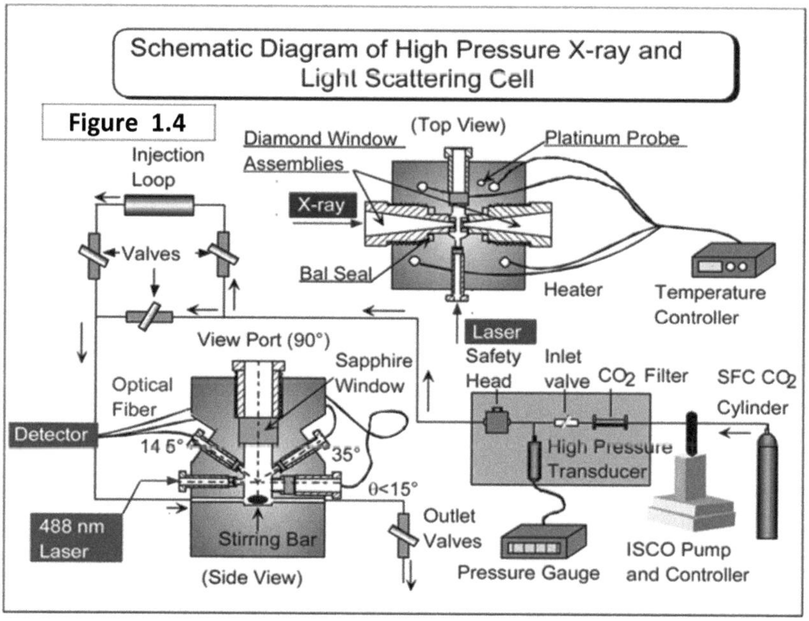

Figure 1.4 A schematic diagram of high-pressure X-ray and light scattering cell.

The reader should pay attention that the scattering cell shows both the top view and the side view so that a better perception on the cell design can be realized. Note the directions on how the X-ray beam and the laser beam pass through the scattering cell and where the light scattering detectors are located. Please also note that in the detector section, the values of 145° and 35° are for the observation of scattered light at those scattering angles. The cell block has a temperature controller with temperature sensors. The current design was being used primarily for studying supercritical CO_2 behavior.

In Figure 1.4, the schematic diagram was not drawn to scale. While the light scattering cell was designed to be fairly small as the high-pressure chamber with thick walls should be fabricated in such an approach that it can be inserted into the synchrotron X-ray beam path. The controls and measurements for high pressure and temperature also take up some space. Note that the viewport with sapphire window permits visual observation of the sample (under high pressures). The system was designed in modular units for easier assembly, disassembly, and transport purposes.

In the lower left quadrant of Figure 1.1, a time (t) evolution of X-ray scattered intensity for both SAXS and WAXS in terms of intensity (I) and scattering vector ($\mathbf{q}$), is presented. Here, we have used WAXD and WAXS interchangeably, as the materials may exhibit small-scale inhomogeneity (WAXS, as stated in Figure 1.1), instead of structures (WAXD), as denoted in the text.

Figure 1.2 shows a rheo–X-ray apparatus which combines rheometry with synchrotron SAXS/WAXD measurements. In rheometry, there are two types: one for oscillatory shear and the other for steady shear. In addition to the schematic figure, there is also a corresponding picture showing how such an instrument looks like. As the apparatus needs to be inserted into the synchrotron X-ray beam path, between the synchrotron X-ray collimation system for the incident X-ray beam and the detection system to measure SAXS/WAXD (or SAXS/WAXS) X-rays from the sample during oscillatory or steady shear, instrumentation space allocation should be considered in advance.

In the Oscillatory Shear diagram as shown in the upper half of Figure 1.2, it is important to have the vacuum pipe for the X-ray beam (as presented in the middle right for a heavy-walled tube to get very close to the sample chamber, located near the center of the diagram). The vacuum cone-shaped window used a thin Kapton film at the large end, as being displayed in the upper middle of the photograph.

The Steady Shear diagram, as shown in the lower half of Figure 1.2, shows how the X-ray beam enters the thin shear cell and then there is a large vacuum cone between the shear cell and the large MAR-CCD area detector, as shown in Figure 1.3, left middle. The picture on the right side of the Steady Shear apparatus shows mainly the sample chamber.

Figure 1.3 shows simultaneous X-ray scattering and Raman scattering for fiber spinning studies. The Raman probe is pointing at the fiber with the Raman spectrometer being located elsewhere. The MAR-CCD is the X-ray area detector, looking at the fiber, with X-rays passing through the spinning unit. Thus, such a setup is more complex because it involves a combination of not only two types of measurement units (X-ray scattering and Raman scattering), but also a fiber spinning unit which has a coagulation bath. On the left side of Figure 1.3, note the very thin line at the end of the arrow, under the word "Fiber", pointing toward a vertical very light "grey" line. That vertical "grey" line represents the thin PBO fiber, which is hardly visible, as a result from changing the color photo to grey scale for Figure 1.3.

Figure 1.4 shows a schematic diagram of the high-pressure X-ray and laser light scattering cell. The reader should pay attention that the scattering cell shows both the top view and the side view so that a better perception on the cell design can be realized. Note the directions on how the X-ray beam and the laser beam pass through the scattering cell and where the light scattering detectors are located. Please also note that, in the detector section, the values of 145° and 35° are for the observation of scattered light at those scattering angles. The cell block has a temperature controller with temperature sensors. The current design was being used primarily for studying supercritical CO_2 behavior.

In Figure 1.4, the schematic diagram was not drawn to scale.

While the light scattering cell was designed to be fairly small as the high-pressure chamber with thick walls should be fabricated in such an approach that it can be inserted into the synchrotron X-ray beam path. The controls and measurements for high pressure and temperature also take up some space. Note that the viewport with a sapphire window permits visual observation of the sample (under high pressures). The system was designed in modular units for easier assembly, disassembly, and transport purposes.

The reason for presenting such a scattering cell, as shown in Figure 1.4, is to remind the reader that in experiments involving multiple parameter controls and reading of the experimental data, it is worthwhile to examine the conditions of the laboratory, the synchrotron facility, the travel distance, and the shipping requirements for instruments to be used in carrying out such an experiment. The reader should also be aware of a variety of instruments, which are often available in synchrotron X-ray facilities. For example, at NSLS-II (with 29 operating and space for up to 60 beam lines), Brookhaven National Laboratory (BNL) requests that the use of instrumentation from the Center for Functional Nano-materials (CFN) and that from NSLS-II can be joined together, permitting multi-facility and multimodal arrangements on a single general user proposal. Furthermore, the development with the NEXT-III project for future new beam line capabilities has been in progress. Thus, users should always pay attention of updates with regard to a beam-line facility.

1.2.3 Research Opportunities via Scattering/Diffraction (SAXS/WAXD) for Advanced Materials

For advanced materials related to macromolecules, there are many topics that can take advantage of synchrotron facilities. A few example themes are listed as follows:

- Changes in supra-molecular structures of biological/biosynthetic macromolecules, ionomers, block copolymers, dendrimers, charged gels and colloidal (nanoparticles: organic/inorganic; different shapes) systems in solutions, suspensions, and melts;
- Kinetics of phase transition in semi-crystalline polymers (crystallization and melting), multicomponent mixtures and molecularly segmented systems (phase separation);
- High-pressure experiments (hydrostatic pressure, super-critical solvents) for environmental process development;
- In situ chemical/physical changes and polymerization reactions;
- Polymer fibers/films during processing (spinning, extrusion, deformation);
- Polymer melts/solutions in flow and flow-induced crystallization.

We have studied the stress-strain relationship and 2D/3D WAXD patterns during the stretching (extension and retraction) of vulcanized natural rubber, as shown in Figure 1.5. It represents another demonstration in strength on time-resolved and in situ measurements.

We can visualize that, in order to accommodate such an instrument into the synchrotron X-ray beam line, space consideration for the sample chamber and integration with the incident X-ray collimation system again need to be carefully considered in advance. Sometimes, one also wishes to use different synchrotron facilities. Then, different space and transportation requirements should best be taken into account in the initial design.

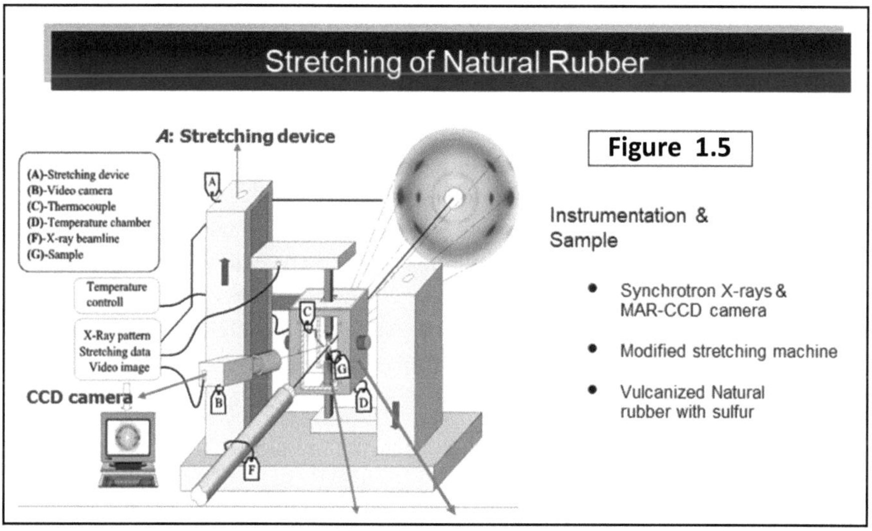

Figure 1.5 Stretching of natural rubber. Note the symbols for (A) stretching device, (B) CCD video camera, (C) small thermocouple, (D) temperature chamber with thick insulating walls, (F) X-ray beam aiming at the sample of interest, and (G) the rubber sample that is being stretched. Finally, the circular figure showing a diffraction pattern on upper right illustrates the results from the MAR-CCD two-dimensional detector (camera).

We can visualize that, in order to accommodate such an instrument into the synchrotron X-ray beam line, space consideration for the sample chamber and integration with the incident X-ray collimation system again need to be carefully considered in advance. Sometimes, one also wishes to use different synchrotron facilities. Then, different space and transportation requirements should best be taken into accounts in the initial design

1.2.4 Brief Remarks on Synchrotron X-Ray Instrumentation

The intention for a brief presentation of the SUNY Beam Line at X27C of NSLS is to illustrate a few features which may be worthy of consideration when one deals with supplementary instrumentation as briefly described in previous sections. New beam lines with more capabilities shall continue to advance at synchrotron facilities. For example, NSLS has been replaced by NSLS-II (with NSLS-III being discussed and planned) at Brookhaven National Laboratory on Long Island, New York. All beam lines want to make best use of the facilities so that the expensive investment for a synchrotron facility can be justified. There are supplementary instructions and teaching tools included with each beam line. The purpose of the current remarks on synchrotron X-ray instrumentation is to show how synchrotron beam lines can accommodate new (none facility) instruments whereby specific desirable features can be taken into consideration. Two features are being raised for further discussions.

The beam line should have a stable platform for convenient introduction of supplementary instruments in connection with synchrotron X-ray beam lines. An example is to use readily available optical tables, as shown in Figure 1.6, where specific desirable features can be taken into consideration. Two features are being raised for further discussions:

(1) The beam line should have a stable platform for convenient introduction of supplementary instruments in connection with synchrotron X-ray beam lines. An example is to use readily available optical tables, as shown in Figure 1.6.

(2) Figure 1.7 shows an amplified *overall side view of the collimation system*. Note the precision demand on the movements of the two pinholes because of the small pinhole sizes for the incident X-rays. The long vacuum tube behind the X-ray collimation system is the passageway for measurements of SAXS and WAXD. In the present case, a MAR-CCD was being used. Figure 1.8 shows *amplified side and top views of the collimation system*. Again, due to the small pinhole sizes of the X-ray collimation system, very-high-precision and stable micrometers were used. It is noted that variations of different dimensions for the overall system are quite flexible because of the stable platform based on the use of optical tables which have standard screw holes on the optical table surface for attachment of instruments of interest.

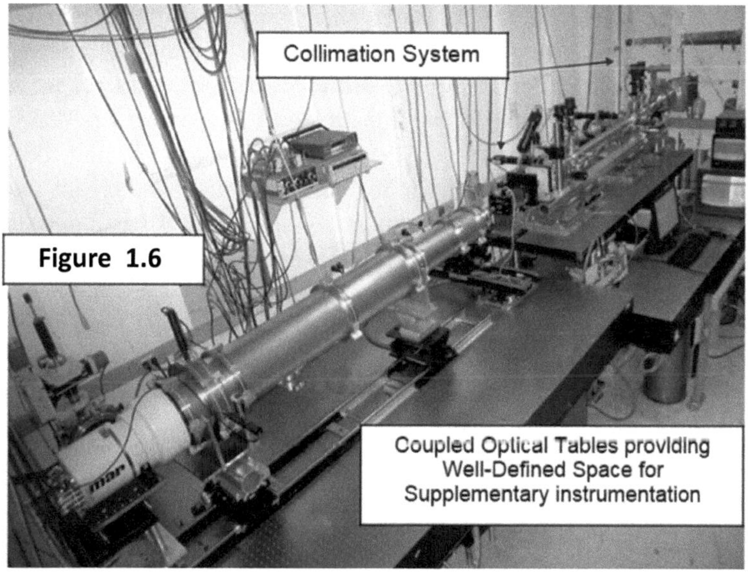

Figure 1.6 Advanced Polymers Beam Line, National Synchrotron Light Source.

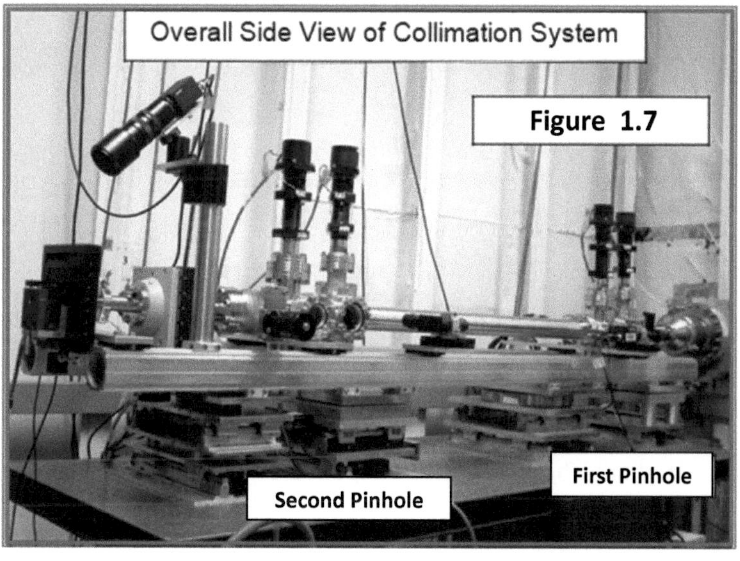

Figure 1.7 Side view of the collimation system. Note the precision demand due to small cross sections of the X-ray beam: high-precision micrometers, stable optical table bases, and consideration of spatial stability, even including the thermal coefficient of expansion of the materials being used and temperature maintenance of the hutch.

The two primary topics of this book were developed with efforts and thoughtful contributions by many members of my research team at Stony Brook over the years, especially those involved in early instrument development by Dr. Walther Tscharnuter and Dr. Bruce Weiner. In particular, I would like to acknowledge my appreciation to Professor Dufei Fang for his work on laser light and synchrotron X-Ray scattering instrumentation. For their contributions, a formal acknowledgement is presented at the end of the book.

In Figure 1.7, the side view of the collimation system shows some arrangements on the three pinhole locations, its optical table foundation from the ground up and the demand on the *stability of pinhole locations from the ground up.* The 4 (black colored) precision micrometers which control the two (x and y) pinhole positions for the incident X-ray beam have controls to the order of microns. The overall positioning of the synchrotron beam line is shown in the figure for Simultaneous Laser Cross-Correlation & Synchrotron SAXS/WAXD in Chapter 2.

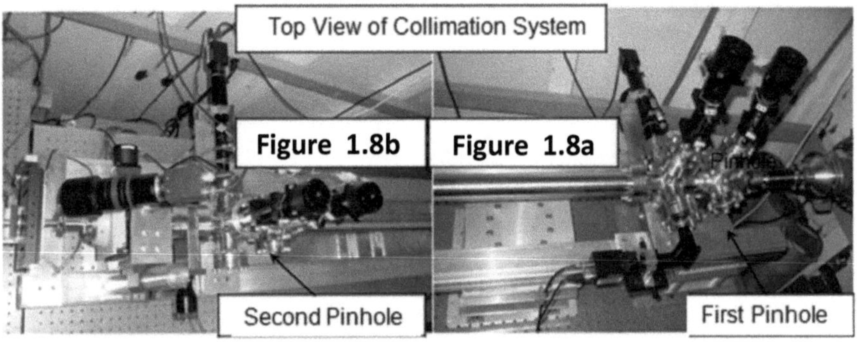

Figure 1.8 Top view of the collimation system as based in Figure 1.7, showing the amplified views of first and second pinholes.

It should be noted that Chapter 2 is physics oriented, while studies on functional soft matter and macromolecular self-aggregation deal with systems related to materials chemistry and synthesis. The crossover is what makes such studies more interesting but also more challenging. Normally, it is not advisable and quite difficult to be an expert on everything. The key advice is for the reader to be able to learn the basic principles behind the chemical or/and physical processes and to get to know who the experts in the field(s) of interest are. Here, we refer to Section 1.1.2, "Notes on Research Adviser."

1.2.5 Brief Note on Photon Cross-Correlation Function in Laser Light Scattering

The short subsection here deals with a brief introduction on Photon Cross-Correlation Function in laser light scattering. It is an advanced topic which is being very briefly presented at this time because of the intention to try to separate instrumentation from discussions on principles. Photon cross-correlation is one of the two main themes for Chapter 2.

A collection of two slides and a short explanation on photon cross-correlation, three-dimensional dynamic light scattering, and normalized cross-correlation function provides the reader a taste of the next chapter, Chapter 2.

We single out the X-ray collimation system as a preview of the instrumentation associated with a SAXS/WAXD beam line in a synchrotron X-ray facility. The instrumentation associated with synchrotron X-ray beam lines can roughly be separated into three sections: the collimation system, the sample chamber and associated additional instrumentation, and the detection system.

1.2.6 Photon Cross-Correlation Function Preliminaries

From instrumentation that can be coupled with synchrotron X-ray facilities, a preliminary introduction is being presented to provide the reader information on the more recent development as related to Chapter 2. These are photon cross-correlation and three-dimensional dynamic light scattering (as presented under dynamic light scattering (DLS) optical schemes in Chapter 2). The fundamental equation relating the normalized cross-correlation function, $g^{(2)}(Q, \tau)$ to the field cross-correlation function $f(Q, \tau)$ is shown in Figure 1.9. More details of photon cross-correlation, 3D-DLS and normalized cross-correlation function will be presented in Chapter 2. Pay attention to the information on the meaning of β, β_T, β_{OV}, and β_{MS}.

Normalized Cross-Correlation Function

Figure 1.9

$$g_C^{(2)}(Q,\tau) \equiv \frac{\langle I_1(Q,0) I_2(Q,\tau)\rangle}{\langle I_1(Q)\rangle\langle I_2(Q)\rangle}$$

$$= 1 + \beta\beta_{OV}\beta_{MS}\beta_T\, |f(Q,\tau)|^2 \qquad \text{with } f(Q,\tau) = \frac{\langle E_{1,2}^S(Q,0) E_{1,2}^{S*}(Q,\tau)\rangle}{\langle I_{1,2}^S(Q)\rangle}$$

- β represents the coherence factor, as defined in single photon correlation.
- Maximum value for the product of all β_{ij} values reveals the signal-to-noise ratio that a given dynamic light scattering experiment can achieve.
- Overlap factor, β_{OV}, takes into account that each detector sees a slightly different scattering volume.
- β_{MS}, is determined by ratios of average intensities of ***single scattering***, $I^S_{1,2}$, and total scattering:

$$\beta_{MS} = \langle I^S_1(Q)\rangle \langle I^S_2(Q)\rangle / [\langle I_1(Q)\rangle \langle I_2(Q)\rangle]$$

- Last factor, β_T, is a technique factor. $\beta_T = 1$ for TCDLS. For 3DDLS, the maximum intercept is *only about half* the value of TCDLS.

Figure 1.9 Normalized cross-correlation function.

In Chapter 1, the intention is not only to provide an overview on this (introductory) book; BC tries to use his personal experience to prepare less experienced students to meet new challenges in our changing world with an optimistic viewpoint. The topic on macromolecular assembly and engineering offers opportunities of benefits to society. The emphasis has been on an overview of the physical techniques of laser light scattering and synchrotron X-ray scattering/diffraction with supplementary instrumentation and their applications to materials research. One specific topic on electro-spinning/blowing technology with applications to the biomedical field as well as water (and air) purifications was single out and presented in Chapter 3 to implicitly demonstrate such themes as an essential element in the well-being of our lives.

1.3 A Final Note

We return to one of the book themes on *Macromolecular Aggregation and Engineering*. While our primary interest includes macromolecular and supra-molecular systems in Chapter 3, it is worthwhile to point out that in applications, especially when dealing with biological systems, our interest has a broader perspective, covering not only macromolecular and supra-molecular systems but also down to the cellular level, dealing with assemblies at molecular dimensions. There is a distinction between macromolecular aggregation and molecular assembly. The reader may be interested in finding out the subtle details of the difference in concepts, as discussed in [2].

In dealing with macromolecular materials, it is important to consider the source, e.g., whether petroleum or plant based. Needless to say, the emphasis should be on plant-based sources where cellulose can play an important role in the development of base components for water (air) filtration systems.

To supplement the initial discussions being presented in Chapter 2, BC wants to make two more additional comments:

The first comment is prompted partially by using (discarded) old mechanical components in order to illustrate the principles of laser light scattering measurements, as has been presented in Chapter 2 (e.g., Figure 2.20 A laser light scattering photometer). This practice is acceptable because there is a fundamental difference in demonstrating fundamental principles and in the judicial use of practical tools. The reader is reminded that in our studies, we should always remember a distinction between function and cost.

The second comment deals with patent number 11596886 on nanostructured fibrous membranes for membrane distillation (Inventors: Benjamin Chu and Benjamin S. Hsiao, with this work being discussed and presented as reference 22 at the end of Chapter 4. Its applications for water/air purifications, as described in Chapters 3 and 4, further suggest to us the inter-relationship of topics presented in this book). The key point in this case is that we can fabricate nano-fibrous membranes by using the electro-spinning process. However, cellulose nano-fibers can also be fabricated by other more economical means.

In Section 1.1.4, on the Future in Research as related to Macromolecular Aggregation and Engineering, we briefly mentioned "climate change," which shall have an impact on our lives. Its broad-range impact is worthy of our attention. We shall now take this opportunity to present some of its profound effects on the well-being of our society. While our main themes of this book deal with Laser Light and Synchrotron X-Ray Scattering (Chapter 2) and Evolution of Electrospinning Technology (Chapter 3), we are concerned with the topic whether Water is Our Next Challenge (p. 131 in Chapter 4). Thus, climate change has a broad effect permeating through the topics of this book.

It should be of interest to know that "water" is a topic related to our energy needs. While we aim for green energy of solar, wind, and nuclear, there is no perfect energy source. For example, the current shift of storm locations in the USA has flooded some states and left other places dry. Then, some sources of water passing through dams are being affected, leaving reservoirs getting dried up and affecting the hydroelectric power of even large dams.

Another interesting example is to take a look at the Ogallala aquifer, under a substantial portion of the Kansas state in the USA. The Kansas corn fields depend largely on water from the aquifer. Climate change has shifted the US rainfall location. While other states are witnessing floods, the Ogallala aquifer is not being replenished as before. Thus, the corn fields in Kansas may be affected by climate change in a profound way in the future.

References

1. Robert Lefkowitz, The Art of Scholarly Mentoring (2021). Available at: https://www.insidehighered.com/advice/2021/01/26/nobel-laureate-shares-10-rules-being-effective-mentor-young-research-scholars (accessed on April 11, 2023). *A Nobel laureate shares 10 rules for being an effective mentor of young research scholars* (opinion).
2. Chi Wu, *Macromolecular Solutions*. Higher Education Press, Beijing, China; 521 p, pp. 476–479 (2021).

Chapter 2

Laser Light and Synchrotron X-Ray Scattering

2.1 Outline

This chapter deals with the techniques of laser light and synchrotron X-ray scattering. The combination of laser light with synchrotron X-rays as the probing electromagnetic radiation can expand the range of studies on the structure and dynamics of many materials, among other topics. In the current context, the applications are mainly concerned with macromolecular and supra-molecular systems over a very broad range of length scales.

The development starts with a general description on electromagnetic radiation from microwave via the visible to ultraviolet, and then going to the X-ray range, where the X-ray radiation, produced by synchrotrons, plays an indispensable role. This chapter starts with laser light scattering (LLS). Nevertheless, the role of X-rays is so very closely related to LLS that the connections are interwoven with LLS in the discussions. The intention is to remind the reader of the importance of this connection, even though historically, the two topics have been developed independently, in essence, by scientists/engineers, often with different backgrounds in their scientific training.

Laser Light and X-Ray Scattering: With Evolution of Electrospinning Technology in Biomedical Applications and Water Purification
Benjamin Chu

ISBN 978-981-4968-88-1 (Hardcover), 978-1-003-54074-8 (eBook)
www.jennystanford.com

The presentation begins with connections among light, X-rays and neutrons (Figures 2.1 to 2.3). Then, an illustration on the use of the Moiré pattern for the crossing of two laser beams is presented (Figures 2.4 to 2.13). In the scattering geometry, basic information on static and dynamic light scattering with discussions on the detection optics, the Siegert relation, and the time correlation function is presented. From here on, the words "Figures" (or Tables) are often being overlooked or dropped.

In transition from laser light toward synchrotron X-rays, as the probing radiation of laser light and synchrotron X-rays has their unique characteristics, it is worthwhile to pay attention to the design of X-ray optics. Furthermore, instrumentation for synchrotron X-rays is mostly different from conventional laboratory X-ray instrumentation. With additional associated instrumentation, as has been presented partially in Chapter 1, permitting measurements of other physical parameters that are useful in studies related to macromolecular structures and dynamics so as to be able to take advantage of synchrotron X-rays. (Section 1.2.4), the concept of photon cross-correlation was introduced (Section 1.2.6), before returning to synchrotron X-ray scattering. The simultaneous use of laser cross-correlation and synchrotron X-ray scattering/diffraction becomes the ultimate aim of the presentation in this chapter.

2.2 Laser Light/Synchrotron X-Ray Scattering

2.2.1 Introduction

In this chapter, the main theme deals with laser light/synchrotron X-ray scattering techniques. The development of light scattering started with the late Professor Peter JW Debye during World War II when there was a rubber shortage for the war effort. Professor Debye designed the first light scattering instrument with the help of a very good machinist at Cornell University and discussed the design when they went fishing together. Ben Chu (BC) first learned the experimental technique from working on that instrument with his postdoctoral colleagues (Dietrich Woermann and Hans Coll), starting in 1959. Over the years, laser light scattering (LLS) has been one of his key research tools.

BC has selected 10 publications (1970–2015) to illustrate his continued interest in this technique· See the author's selected light scattering book/review publications [1–10].

Light scattering started before the invention of the laser. By changing the light source from a conventional light source, such as a mercury arc lamp, to a coherent laser source, remarkable effects on its applications to dynamic studies of macromolecular systems have taken place. Measurements of dynamic properties on systems of interest now become feasible. At about the same time, it has become known that many of the concepts already developed in World War II for microwave detection in aviation can be transferred from wavelengths used in the microwave range to the shorter visible wavelength range, eventually across the electromagnetic wave spectrum from infrared through visible and ultraviolet wavelengths, finally reaching the X-ray range by means of modern-day X-ray synchrotrons.

At this stage, one should consider that the learning of new discoveries is a pleasure in life. The student reader has *no age limit*, i.e., we accept the learning process throughout our lives. It is worthwhile to note, based on past experience over the years, that when seeking advice, the student should try to remember the attitude based on *mutual* respect and interest. As has been mentioned in Chapter 1, the more pertinent and fruitful discussions, especially for beginning graduate students in science or/and engineering, are often related to impacts on science/engineering and of benefits to society.

We shall start with some terminology on the current topics so that the discussions on the fundamentals of laser light and X-ray scattering can be carried out smoothly. In this section, we first provide a brief outline of some of the steps (a total of 7 steps) as follows:

1. Moiré patterns on crossing of two laser light beams: Sections 2.2.1 to 2.2.10
2. Conservation of *momentum* and *energy* in LLS (see discussions in reference books, e.g., #8 in author-selected light scattering books), Section 2.3.1 and Figure 2.14
3. Fundamentals on static light scattering (SLS) and dynamic light scattering (DLS): Sections 2.3 to 2.5

4. Synchrotron X-rays: Section 2.6
5. Associated instrumentation for synchrotron X-ray experiments: Section 2.6.5
6. Photon cross-correlation: Sections 2.6.6 and 2.6.7
7. Simultaneous photon cross-correlation and synchrotron X-ray scattering: Sections 2.6.7

Before we start with those 7 steps, we present some background information, starting with the schematic representation on the scattering vector **q**, as shown in Figure 2.1:

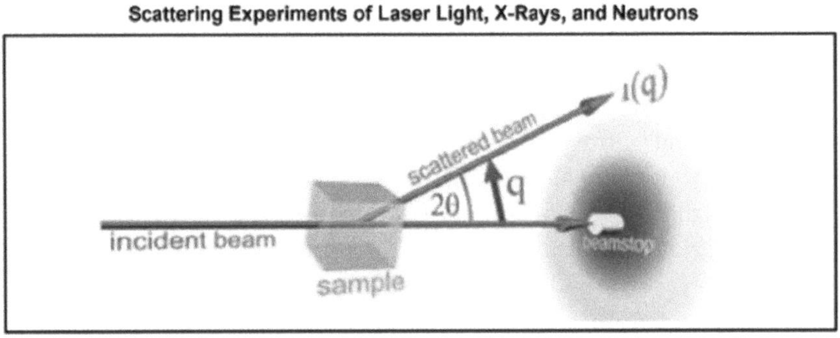

Figure 2.1 Relationship between the scattering vector **q** and the real space *d* with $d = 2\pi q$.

The scattering vector **q** with an absolute value of $q = |\mathbf{q}| = (4\pi/\lambda) \bullet \sin\theta$, where λ is the wavelength and θ is *one half of the scattering angle*. The reciprocal relationship between the Bragg spacing, *d*, in real space and the scattering vector, **q**, in reciprocal space is $d = 2\pi/q$. The reader should remember the change in the definition for the scattering angle (whether θ or 2θ) which often occurs in the literature.

Before we start with the discussions on light/X-ray scattering techniques, a brief overall perspective of the light/X-ray (and neutron) scattering with other complimentary techniques is provided for the reader. Figure 2.2 shows the connections among light, X-rays and neutrons, including relationships among both static and dynamic properties on the size range, e.g., from Angstroms to microns and dynamic properties with time scales covering a very large range, where PCS is photon correlation spectroscopy; TOA denotes the "time-of-arrival" technique; and FP is associated with *fluorescence photo-bleaching*. The scattering techniques can also be associated with system specific methods, such as fluorescence recovery after photo-bleaching (FRAP) where labeling of species is required or electrophoretic light scattering where movements of charged species are involved.

	Light Scattering	**SAXS**	**SANS**
	Static Properties: Spatial Correlations		
λ(Å)	~3000	~1.5	~5
***I*(*q*))**	Large scale	Small scale	
	Dynamic Properties: Time Correlations		
q^{-1}	~100 nm - μm	1 nm – 100 nm	
I (*q*,*ω*) , *sec*	$1\text{-}10^{-6}$ *(PCS)* 10^{-8} *(TOA)* 10^{-12} *(FP)*	($\sim10^{-3}$) Fast *(spin echo)*	
	System Specificity		
Labeling	Fluorescence Recovery after Photo-bleaching Forced Rayleigh Scattering	Anomalous Scattering	H,D Substitution
Charge	Electrophoretic Light Scattering		

Figure 2.2 Connections among light, X-rays, and neutrons; H, D denote hydrogen, deuterium.

There is an interesting relationship between frequency and wave vector for laser light and X-ray scattering when in comparison with a range of other spectroscopic techniques, including Raman scattering, Brillouin scattering, and X-ray photon correlation spectroscopy (XPCS). The key point is to note that, with current synchrotron X-ray sources of unprecedented brilliance, e.g., of $\sim10^{21}$ photons/sec/mrad^2/mm^2/0.1% bandwidth for X-ray photons at NSLS, measurements can be extended toward faster time and shorter length scales, thereby approaching the neutron spin-echo technique as well as closing the gap in phase space between XPCS and free electron laser (FEL) experiments.

We shall start with an overall concept on light scattering spectroscopy, as presented by Figure 2.3. The laser provides the incident light (I_0). However, its light signal can be modified by another light signal or time-dependent signal, e.g., electric signal, arbitrarily denoted as I_t. The incident radiation impinges upon the sample (as represented by green colored dots in the colored slide format), with the transmitted beam being captured by a Rayleigh horn, which is being used to avoid back transmitted light from hitting the sample, and thereby yielding a scattering process in the opposite direction. The scattered light (I_s) is detected by a detector, as represented by a PMT (photomultiplier tube). The reader can use this figure to gauge the level of knowledge on laser light scattering. On first exposure to the topic, as often is the case for a beginning chemistry graduate student, the reader should start consulting with a light scattering book, as shown in the author's selected light scattering book/review publications [1–10] (publications #2, #8, and #9), or another good book, for examples, by Bruce J. Berne at Columbia University and Robert Pecora at Stanford University on dynamic light scattering or (in parts) by Michael Rubinstein now at Duke University and Ralph H. Colby at Penn State on polymer physics.

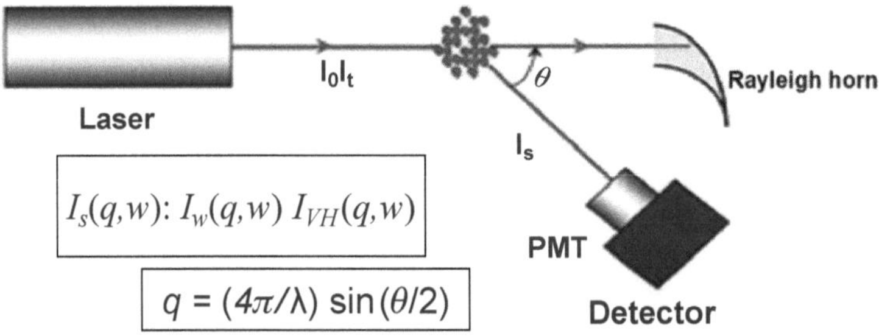

Figure 2.3 Laser Light Scattering. θ is the scattering angle; subscript *VV* denotes vertically polarized incident and scattered light; subscript *VH* denotes vertically polarized incident and horizontally polarized scattered light. Note the difference in definition for θ in Figures 2.1 and 2.3.

The primary aims of this section are to emphasize the physics, instead of presenting mathematical equations, in order for the reader to get a better sense (1) on how laser light scattering works by using Moiré interference patterns and (2) on the basics of laser light scattering so as to be able to understand pertinent experimental requirements in laser light scattering spectroscopy with two-crossed laser beams.

Figure 2.3 shows the essential terms associated with laser light scattering spectroscopy.

In the first step, Moiré patterns on crossing of two laser light beams, or how laser light scattering works by using Moiré interference patterns: as discussed in Sections 2.2.1 to 2.2.10:

Figures 2.4 to 2.13 show the *interference patterns from the crossing of two laser beams.* The primary purpose of these figures *is to demonstrate the important features on laser light scattering spectroscopy when carrying out such experiments.* A grey-scale title format (dark background with white lettering) is being used to separate this set of discussions from other subsections in Chapter 2. It should be noted that the same title format will be used in other sections of importance.

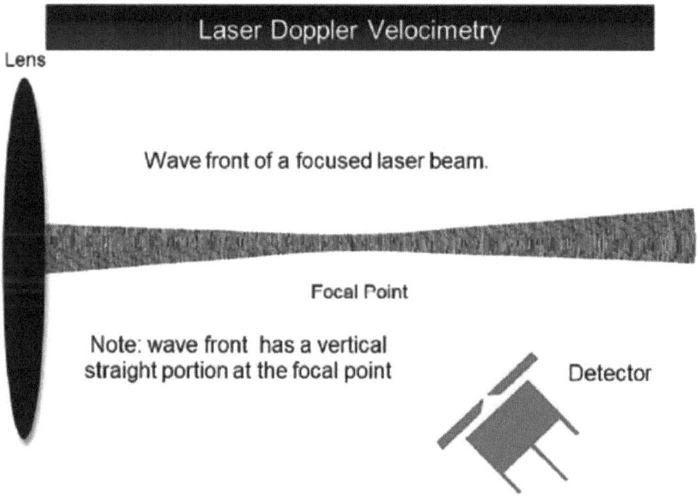

Figure 2.4 Laser Doppler Velocimetry.

On laser Doppler velocimetry, Figure 2.4 shows the wave front of a focused laser beam. The vertical dark lines represent the wave front of a focused laser beam passing through the focal point where the wave front has a limited range of relatively straight vertical lines.

A Moiré pattern of the enlarged interference pattern is shown in Figure 2.5 for the crossing of two laser beams with constructive and destructive interference regions.

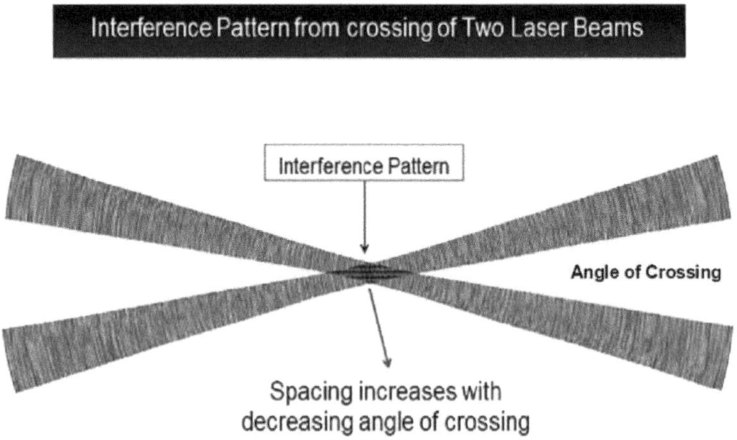

Figure 2.5 Interference pattern from crossing of two laser beams, centered at the focal point of both beams.

As shown in Figures 2.5 and 2.6, it may be of interest to know that the shape of the interference pattern has changed with the amount of angle of crossing. It should be noted that the spacing of the interference pattern increases with the decreasing angle of crossing.

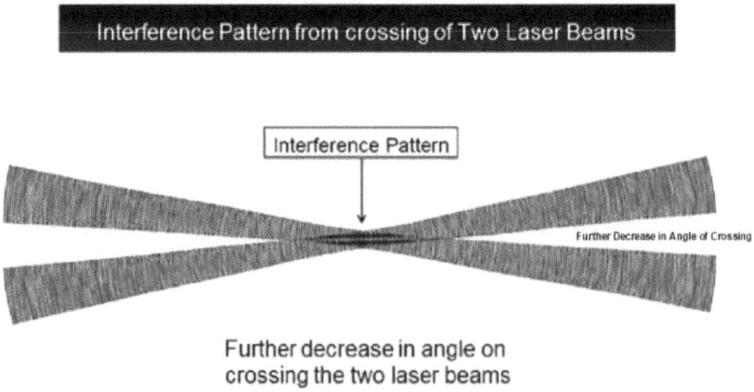

Figure 2.6 Interference pattern from crossing of two laser beams with further decrease in the angle of crossing at the beam waist. Note the increase in distance between constructive and destructive interference.

We may wish to pay attention to the alignment procedures in laser light scattering. By considering different forms of misalignments in the experimental setup a step further, we can realize the experimental uncertainties being introduced in various forms of misalignments. Thus, the lesson here is trying to caution the reader to take a closer look at some of the experimental details in order to have a reasonable idea in carrying out the instrument alignment for laser light scattering experiments.

Figure 2.6 shows a further decrease in the angle of crossing of two laser beams. It should be noted that if the crossing of the two laser beams is not at the waist of the straight portion of the wave front, the interference pattern will produce an uneven spacing, as shown in Figure 2.7.

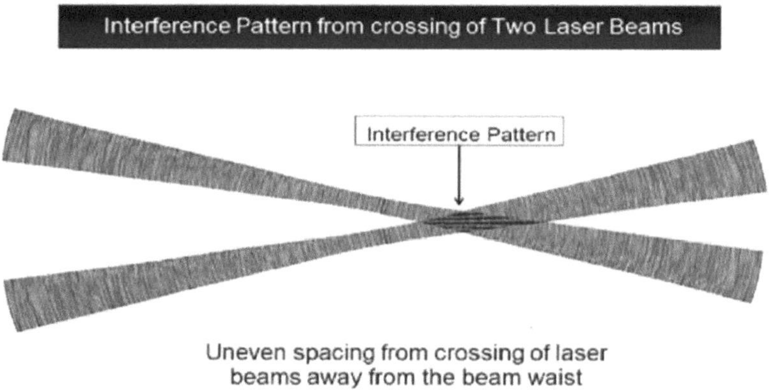

Figure 2.7 Interference pattern from crossing of two laser beams, away from the beam waist (or focal point) of both beams.

Thus, when a particle passes through such an interference pattern, the relationship of d versus t, as shown in Figure 2.8, is no longer valid. The geometrical arrangements in having the two laser beams crossing at the beam waist and in having the particle passing through the crossed laser beams with the direction being perpendicular to the laser beam interference pattern mean that in the experimental setup and alignment procedure, it is advisable to pay attention on the design, precision requirement, and measurement needs for such a setup.

Figure 2.9 shows a setup as related to *laser Doppler velocimetry* (LDV), also known as laser Doppler anemometry which has been well developed, with information available in the literature.

However, LDV is often relatively new for beginning graduate students in chemistry or biology.

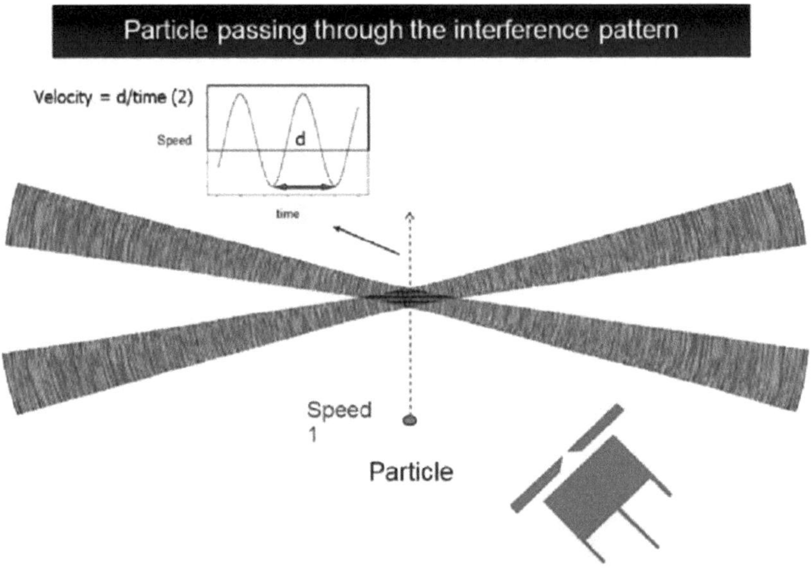

Figure 2.8 Particle passing through the interference pattern (perpendicular to the interference pattern as shown.

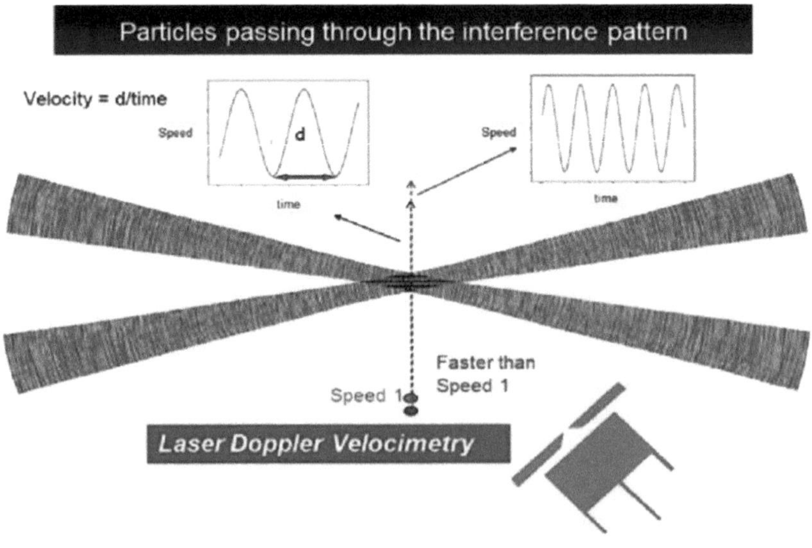

Figure 2.9 Illustration of two individual particles passing through the interference pattern, produced by the two crossed laser beams.

With the brief discussions above, we can now proceed to provide a visual explanation on two basic techniques related to laser light scattering: known as *homodyne* and *heterodyne* techniques. Let us first consider the production of circular waves from the scattering of a polarized particle, as shown in Figure 2.10. When the incoming incident laser beam with frequency ν_1 impinges on the polarized particle, a circular wave is propagated outward from the particle. In the homodyne technique, as shown in Figure 2.11, another laser beam with a local oscillator having a frequency $\nu = \nu_{loc}$ is introduced. We can then see the interference effect between the waves from the particle and those from the local oscillator. For the heterodyne technique, we note that $\nu - \nu_{loc} \pm 6$ V (with the value of 6 V being an arbitrary setting), indicating that, in addition to the local oscillator, an additional displacement in the local oscillator is being introduced, as shown in Figure 2.12.

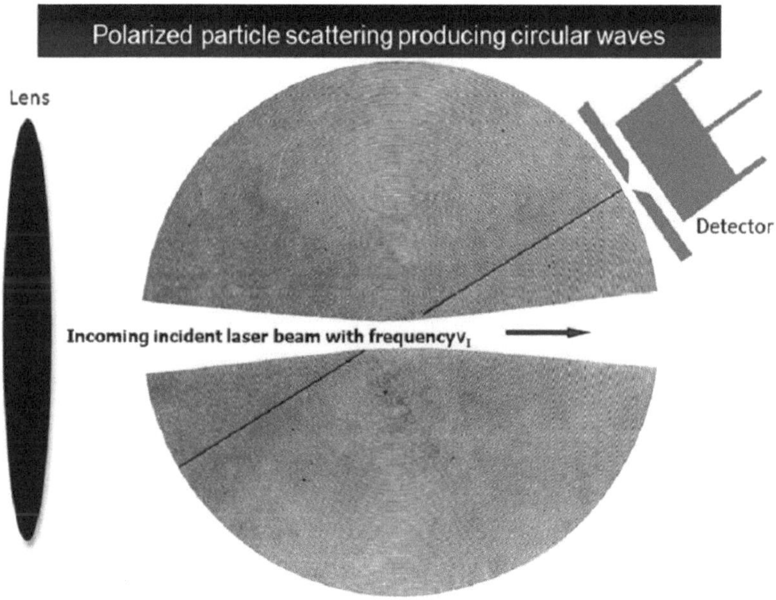

Figure 2.10 Polarized particle scattering producing circular waves which has not yet included the interference with the incoming incident laser beam.

Figures 2.11 and 2.12 show the interference patterns of homodyne and heterodyne techniques.

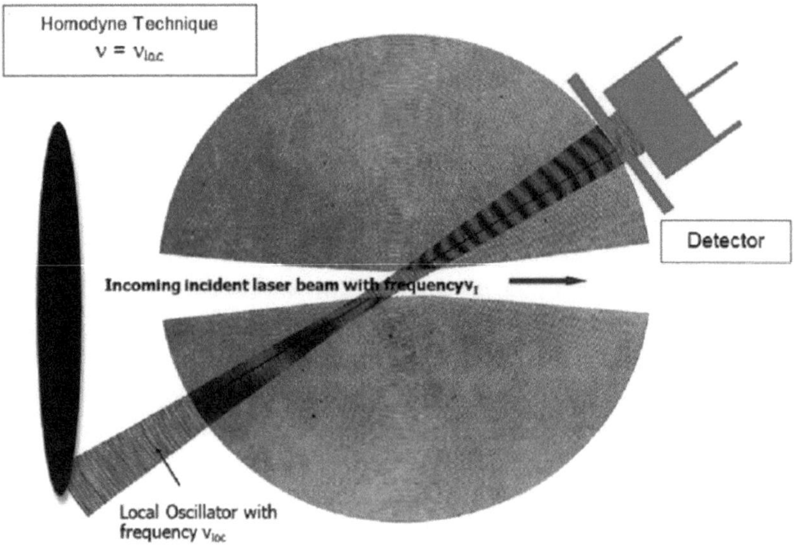

Figure 2.11 Homodyne technique.

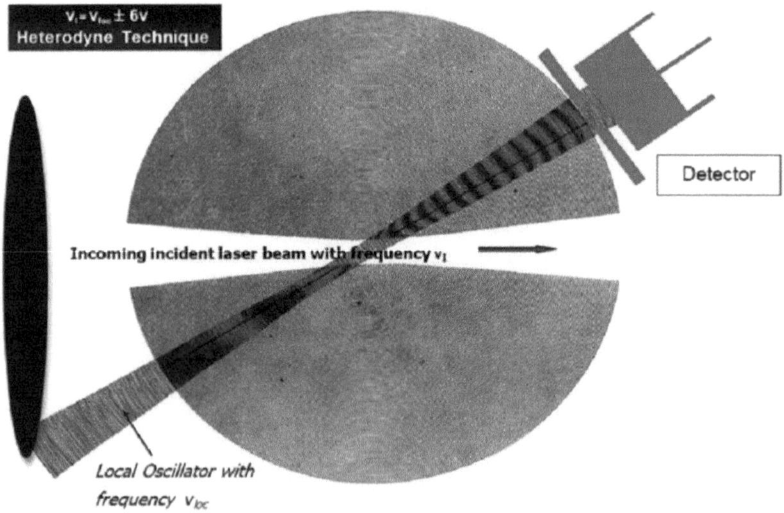

Figure 2.12 Heterodyne technique.

In Figure 2.13, the idea here is to show that many particles in the incident laser beam emit circular waves that can also interfere with one another. Thus, we have static light scattering (SLS) and also dynamic light scattering (DLS) in the *self-beating* mode. In addition to the homodyne and heterodyne techniques, as shown in Figures 2.11 and 2.12, dynamic light scattering can also be performed in the self-beating mode. More details will be presented in later sections, as shown in the Basics of Laser Light Scattering (Section 2.3).

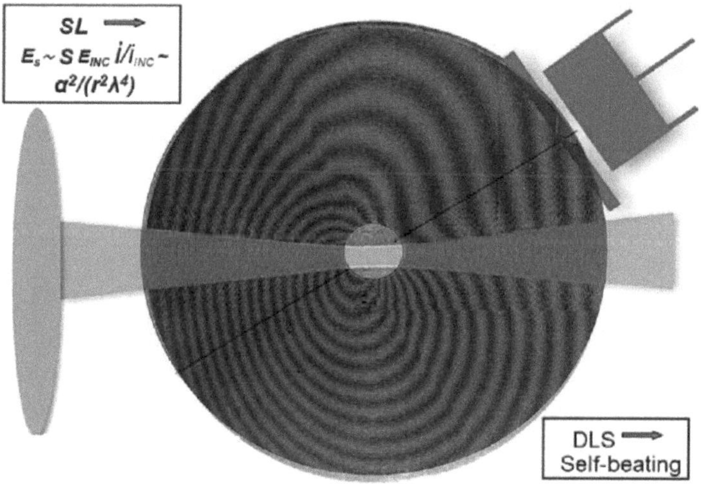

Figure 2.13 DLS in self-beating mode.

2.3 Basics of Laser Light Scattering

In a light scattering experiment, let us first define the scattering geometry and the conditions associated with the conservation of momentum and of energy as shown in Figure 2.14 and discussions shown immediately below.

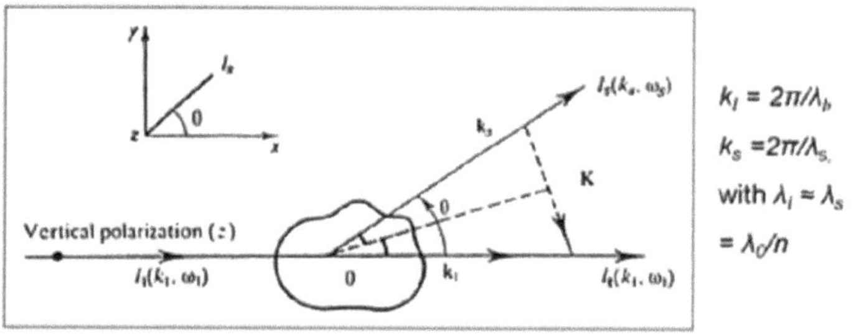

K [= q = Q = (4π/λ)·sin (θ/2)] is the magnitude of the momentum transfer vector.

Figure 2.14 Scattering geometry.

2.3.1 Conservation of Momentum and of Energy

Conservation of momentum:

$$\eta \mathbf{q} = \eta(\mathbf{K}_i - \mathbf{K}_s)$$

Conservation of energy:

$$\eta \Delta\omega = \eta(\omega_i - \omega_s)$$

Where $\eta = h/(2\pi)$ with h being Planck's constant. $\mathbf{K}_i$, $\mathbf{K}_s$, and ω_i, ω_S are, respectively, the incident and scattered wave vectors, $2\pi/\lambda_i$, $2\pi/\lambda_s$, with angular frequencies (ω_I, ω_S) of incident and scattered waves: $\omega_I = 2\pi v_i$, $\omega_S = 2\pi v_S$.

Now, we pay special attention on the coordinate system by setting z in the vertical direction in Figure 2.14. We let the incident light having a vertical polarization so that the scattering is on the yz plane, as shown in Figure 2.14. The magnitude of the momentum transfer vector, which has symbols in K, q, or Q, is defined as $(4\pi/\lambda) \bullet \sin(\theta/2)$, with the conservation of momentum and of energy as

$$\eta \mathbf{q} = \eta(\mathbf{K}_i - \mathbf{K}_s) \tag{2.1}$$

$$\eta \Delta\omega = \eta(\omega_i - \omega_s) \tag{2.2}$$

The normalized electric field time auto-correlation function ($g^{(1)}(\mathbf{q},\tau)$) yields the desired information on the dynamics of the scattering medium:

$$g^{(1)}(\mathbf{q},\tau) = \frac{\langle E(\mathbf{q},0)E^*(\mathbf{q},\tau)\rangle}{\langle E(\mathbf{q},0)E^*(\mathbf{q},0)\rangle} \tag{2.3}$$

What we can measure is the time correlation function of the scattered *light intensity*, which is related to the scattered *electric field* by the relation:

$$\begin{aligned} G^{(2)}(\mathbf{q},\tau) &= \langle I(\mathbf{q},0)I(\mathbf{q},\tau)\rangle \\ &= \langle E(\mathbf{q},0)E^*(\mathbf{q},0)E(\mathbf{q},\tau)E^*(\mathbf{q},\tau)\rangle \end{aligned} \tag{2.4}$$

The normalized time correlation function of scattered light intensity can be expressed as

$$g^{(2)}(\mathbf{q},\tau)=\frac{\langle I(\mathbf{q},0)I(\mathbf{q},\tau)\rangle}{\langle I(\mathbf{q})\rangle^2}$$
$$=\frac{\langle E(\mathbf{q},0)E^*(\mathbf{q},0)E(\mathbf{q},\tau)E^*(\mathbf{q},\tau)\rangle}{\langle E(\mathbf{q})E^*(\mathbf{q})\rangle^2} \quad (2.5)$$

Noting that $\langle I(\mathbf{q})\rangle^2 = \langle I(\mathbf{q},0)I(\mathbf{q},\tau)\rangle$, we then have $g^{(2)}(\mathbf{q},\infty) = 1$.

It should be noted that what we have presented above is to simply write down Equations (2.3) to (2.5) and to ask the readers whether they have come across those expressions before. The *details* are available in light scattering books as have been mentioned earlier.

Our objective for the next development deals with complex time correlation functions associated with crossed laser beams. In order to simplify the discussions, we proceed with an outline of expressions used for standard (single laser beam) laser light scattering experiments. Thus, pertinent equations related to static light scattering, Rayleigh ratios, etc. are first sketched as follows.

2.4 Static Light Scattering

For static light scattering in dilute solution, we have

$$\frac{I_S}{I_I}=\left(\frac{2\pi}{\lambda_0}\right)^4\frac{\alpha^2}{R^2}\frac{N_P}{V} \quad (2.6)$$

Here, each particle is assumed to be a sphere having polarizability α (in vacuo), with N_P/V being the number of particles in volume V. In the case of single particles in a dielectric solvent medium, the polarizability should be replaced by the *excess* polarizability, $\alpha_{ex} = (2\pi n_0)^{-1}(\partial n/\partial C)\ M/N_A$. Then we have

$$\frac{I_{ex}}{I_I}=\frac{4\pi^2}{\lambda_0^4}\frac{n_0^2}{R^2}\left(\frac{\partial n}{\partial c}\right)^2\frac{CM}{N_A} \quad (2.7)$$

where n_0, C, M, and N_A are, respectively, the refractive index of the solvent, solute concentration in mass/volume, particle molar mass, and Avogadro's constant.

The coordinate system for a general scattering scheme is shown below:

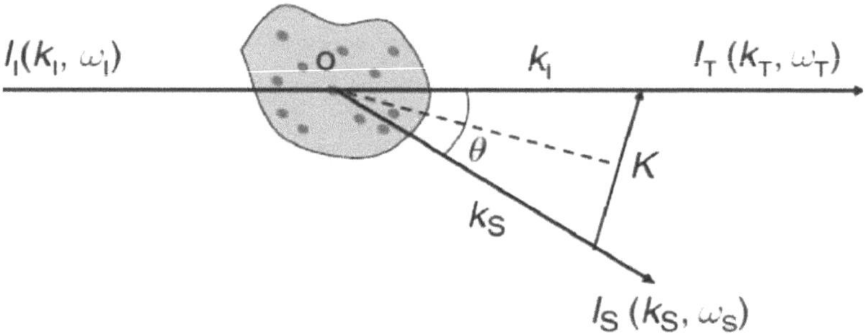

Figure 2.15 Coordinate system for a general scattering scheme.

The symbols have been defined in Figure 2.14, except that k_S is now pointing downward. It should be noted that the scattering plane (xy) is perpendicular to O which is pointing in the z direction.

2.5 Rayleigh Ratio and Excess Rayleigh Ratio

We start with equations for SLS for dilute solution. The Rayleigh ratio R_a and the excess Rayleigh ratio, $R_{a,ex}$, are

$$R_a = R^2 I(\mathbf{q})/I_I \tag{2.8}$$

$$R_{a,ex} = HCM \tag{2.9}$$

with the optical constant H defined as

$$H = \frac{4\pi^2 n_0^2}{\lambda_0^4 N_A}\left(\frac{\partial n}{\partial C}\right)^2 \tag{2.10}$$

The treatment outlined so far applies to solutions of particles. A more generalized approach is needed for condensed systems like complex fluids where the scattered light intensity is determined by fluctuations of the local polarizability, $\delta\alpha(r)$, at position r, in the form of a Fourier transformation of the spatial correlation of $\delta\alpha$.

Without going through further details, we now provide the expression for particles with larger sizes, in comparison with the wavelength of the probing radiation. Then, we have Equation 2.11, where $P(\mathbf{q})$ is the particle scattering factor with molecular interactions being expressed as the second virial coefficient A_2. By considering the first order approximation for $P(\mathbf{q})$, we get Equation 2.12.

$$\frac{HC}{R_{\text{a,ex}}} = \frac{1}{MP(\mathbf{q})} + 2A_2C \tag{2.11}$$

$$\frac{HC}{R_{\text{a,ex}}} = \frac{1}{M}\left(1 + \frac{R_g^2\mathbf{q}^2}{3}\right) + 2A_2C \tag{2.12}$$

In summary, from static light scattering for dilute solutions, we can obtain the molecular weight, M, the particle scattering factor $P(\mathbf{q})$, the radius of gyration R_g, and the second virial coefficient A_2. The coefficient A_2 shows pairwise (two-body) interactions of particles in solution, with the A_2 value being positive for repulsive interactions and negative for attractive interactions. It is interesting to note that to measure those values of M, A_2, and R_g, we need only a simple laser light scattering instrument, as to be shown in Section 2.5.3, **Laser Light Scattering**. In fact, the details are more involved, as we shall discuss after the presentation of a picture of the simple laser light scattering instrument, Figure 2.20. To complete the review on scattering experiments, a few basic expressions on the time correlation functions are described as shown below.

It is the relationship between the *intensity-intensity* time correlation function of a polymer solution ($G^{(2)}(q, t)$) and the *field* time correlation function of the solute ($|g^{(1)}{}_{\text{solute}}(q, t)|$) in a dilute polymer solution. Furthermore, we should consider the experimental conditions as being carried out in the absence of multiple scattering. This experimental restriction can be achieved by designing experiments in order to *minimize* multiple-scattering effects, for examples, by decreasing the beam path length in the sample, by using fiber-optic probes in order to reduce the beam path length, especially in the back-scattering mode, or/and by refractive index matching of the solute (the dispersed phase) and the solvent (the dispersing medium).

The main objective of the section below is to provide some basics on the detection optics. The intention is to let the reader have a basic understanding on how laser light scattering experiments can be carried out to measure some fundamental parameters which shall yield useful information on macromolecules or/and their assemblies of interest. It should be noted that the effects from misalignment can yield uncertainties on the measured parameters and proper instrumentation alignment is a necessary practice for good experiments.

$$G^{(2)}(q,t)=\langle I(q,0)I(q,t)\rangle = A(1+\beta \| g^{(1)}(q,t)\|^2)$$

For macromolecular solution,

$$G^{(2)}(q,t)=\langle [I_{\text{solvent}}(q,0)+I_{\text{solute}}(q,0)][I_{\text{solvent}}(q,t)+I_{\text{solute}}(q,t)]\rangle$$

Then,

$$G^{(2)}(q,t)=A\left\{1+\beta\left[\frac{I_{\text{solvent}}}{I_{\text{solution}}}\left|g^{(1)}_{\text{solvent}}(q,t)\right|+\frac{I_{\text{solute}}}{I_{\text{solution}}}\left|g^{(1)}_{\text{solute}}(q,t)\right|\right]^2\right\}$$

By assuming that light scattered by solvent molecules and particles being not correlated, all cross terms have been dropped. By excluding very short delay time periods,

$$G^{(2)}(q,t)\cong A\left[1+\beta\left(\frac{I_{\text{solute}}}{I_{\text{solution}}}\right)^2\left|g^{(1)}_{\text{solute}}(q,t)\right|^2\right]$$
$$=A\left[1+\beta_{\text{app}}\left|g^{(1)}_{\text{solute}}(q,t)\right|^2\right]$$

It is noted that $\beta_{\text{app}} = \beta(I_{\text{solute}}/ I_{\text{solution}})^2$.

To start with, we introduce two simplest configurations on the detection optics, as shown below. As shown in Figure 2.16, A and B are the two key figures that detail the two primary schemes for the detection optics.

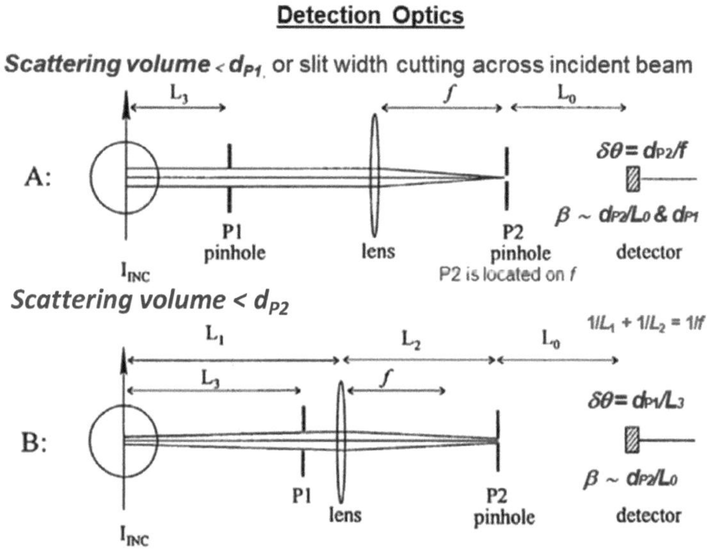

Figure 2.16 Detection Optics. *Note*: The scattering volume is determined by how we use the lens and the two pinhole combinations. The pertinent relationship among the three parameters ($\delta\theta$, β, and the scattering volume) is denoted in heavy characters in grey scale (or blue color) in the figure. The relationship for L_1, L_2, and f is governed by $1/L_1 + 1/L_2 = 1/f$.

2.5.1 Description of Detection Optics for Configurations A and B

In the A-configuration, as shown in Figure 2.17, the scattering volume is mainly determined by the pinhole diameter (d_{P1} ~ 200–400 μm) of P1 (or the slit width for a slit system) that cuts across the incident beam (I_{INC}). The first pinhole (P1) should be as close as possible to the cylindrical scattering cell (being depicted as a circle on the left in Figures 2.16 and 2.17) as is feasible, so that the scattering volume will be better defined. However, due to the existence of the cell holder and the refractive index-matching bath (See a picture for a typical light scattering setup, as shown in Section 2.5.3, **Laser Light Scattering**, the first pinhole is normally 10–15 cm away from the scattering center which is located at the center of the cylindrical cell.

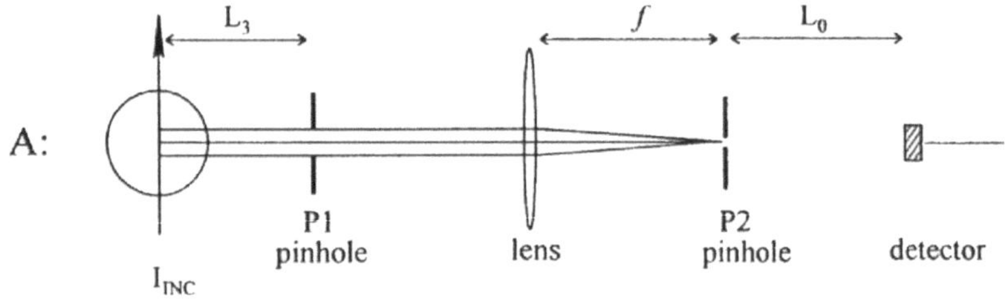

Figure 2.17 Configuration A.

The second pinhole (P2) which is located *exactly* on the focal plane of the lens and the opening angle of d_{P2}/f (with, for example, d_{P2} ~ 100–200 μm and f ~ 10 cm) determines the uncertainty of the scattering angle ($d\theta$). The coherence factor (β) is mainly determined by the opening angle of d_{P2}/L_0 (with L_0 ~ 10 cm) *) but is also influenced by dP1. In this design, the alignment will be easier and the distance between the scattering cell and the detector could be very small.*

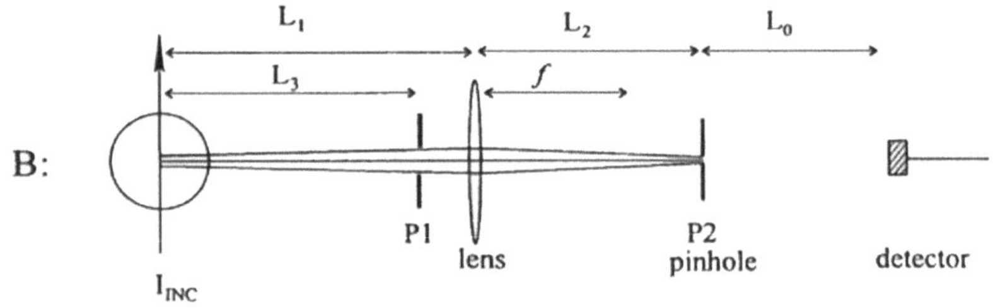

Figure 2.18 Configuration B.

In the B-configuration, as shown in Figure 2.18, L_1, L_2 and f are related by (in heavy characters in Figure 2.16 on **Detection Optics**) with

$$1/L_1 + 1/L_2 = 1/f,$$

i.e., the scattering center and the second pinhole are located exactly on the imagining planes of the lens. The scattering volume is determined by d_{P2} and $\delta\theta$ is determined by the opening angle of d_{P1}/L_3. Therefore, the *first pinhole should be placed as close as possible to the lens,* which has no difficulty in practice. If $L_1 = L_2 = 2f$, one will have the so-called $2f$–$2f$ configuration, i.e., one *optically* moves the second pinhole (P2) to the scattering center. The coherence factor β is still determined by the opening angle of d_{P2}/L_0. In this configuration, f (~5–10 cm) can be smaller than f in the A-configuration. For a given d_{P2}, the scattering volume can be enlarged by a factor of L_1/L_2 (*normally* $L_1/L_2 \sim 3$), so that one can simultaneously have a higher scattered intensity and a smaller $\delta\theta$ for a given distance between the scattering center and the detector. However, a proper alignment of the lens is more difficult to achieve (when compared with the A-configuration) but is manageable. Thus, this design (B-configuration) is often preferred. Please remember that the actual values for the pinhole sizes and the distances, as presented here, are examples.

2.5.2 Siegert Relation and Time Correlation Functions

$G^{(2)}(q, t)$ via the Siegert relation: $G^{(2)}(q, t) = \langle I(q, 0)I(q, t)\rangle = A(1+\beta|g^{(1)}(q, t)|^2)$

A ($\equiv\langle I(q, 0)\, I(q, 0)\rangle$ shown as a (green) horizontal dashed line denotes the baseline, β is a coherence factor. The angled bracket $\langle\ldots\rangle$ denotes an ensemble average.

$|g^1(q, t)|$ $(= \langle E(q, 0)E^*(q, t)\rangle/\langle E(q, 0)E^*(q, 0)\rangle)$ is the *electric field* time correlation function and

$|G^2(q, t)|$ $(= \langle I(q, 0)I(q, t)\rangle/\langle I(q,0)I(q,0)\rangle)$ is the *intensity-intensity* time correlation function.

In our measurements of the intensity-intensity time correlation function, $G^2(q, t)$, the measured values are represented by the solid curve which can approach the base line A (horizontal dash line r, below the solid curve in Figure 2.19); but not the extrapolated short time limit $(A + A\beta)$. Thus, there is an extrapolation of experimental data involved in reaching the $(A + A\beta)$ value.

In the Siegert relation, $\beta = 1$, representing the idealized condition of perfect coherence.

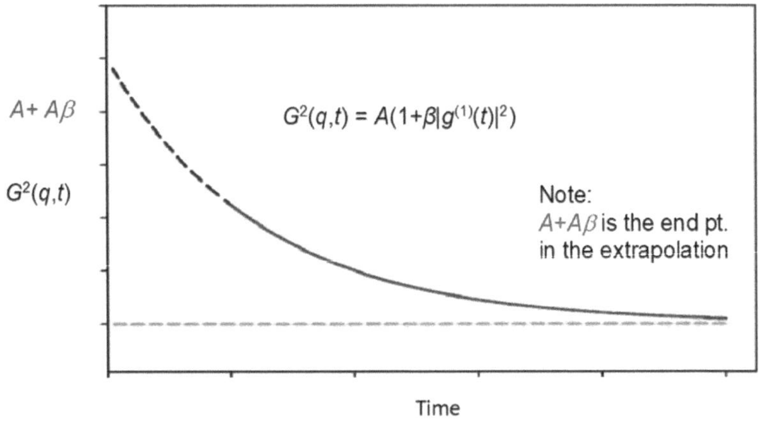

Figure 2.19 Plot of $G^2(q, t)$ as a function of time.

2.5.3 Laser Light Scattering (LLS): Basic Elements of LLS Photometer

In the following figure (photo), the incident laser light passes through a polarizer to select, normally, either vertically or horizontally polarized light. The iris (P1: pinhole 1) is used to define the beam cross section and to clean up the beam edges. The focusing lens follows the configurations, A or B, as shown in **Detection Optics**, Figure 2.16.

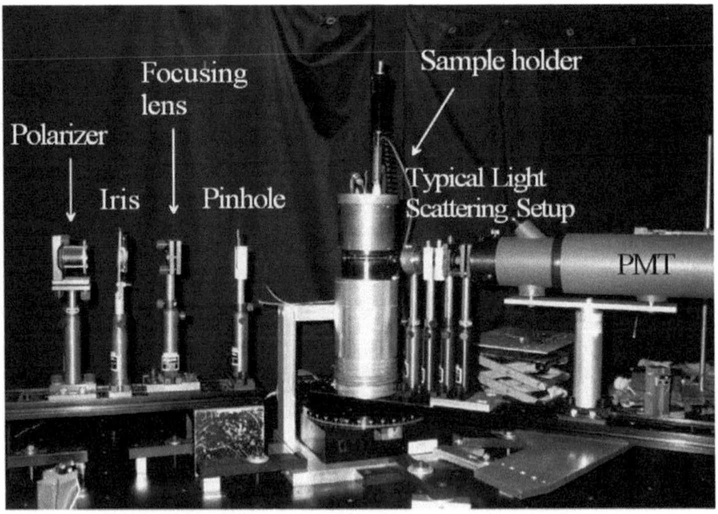

Figure 2.20 A laser light scattering photometer.

It should be of interest for the reader to realize that the use of such a laser light scattering instrument which can carry out both static and dynamic light scattering measurements has hidden details. While the incident beam section shows *polarizer, iris, focusing lens, and pinhole,* the light source is a laser (*not* shown in Figure 2.20) which has already been set to point at the sample of interest. Moreover, *the laser is the built-in reference used to calibrate the instrument.* Note that those mechanical components (as shown in the picture above), while having the capabilities for making rough adjustments, they depend on the laser as the alignment tool. Thus, there is a difference between the degree of freedom needed for the alignment and the tool needed to achieve the alignment, as we have discussed in earlier sections, such as **Detection Optics** in Figure 2.16. Furthermore, Figures 1.7 and 1.8a and 1.8b show the overall side view of the synchrotron X-ray collimation system where the X-ray beam is being aligned and defined by pinholes. The design of the X-ray collimation system is quite different (when in comparison with that of a light scattering instrument) and precision mechanical components are being used to define the incident X-ray beam. In the current X-ray case, no built-in laser was available as an alignment tool.

2.5.4 Static Light Scattering

$$\left[\frac{HC}{R_{\text{vv}}}\right]=\left(\frac{1}{M_{\text{w}}}\right)\left[1+\frac{1}{3}R_{\text{g}}^{2}q^{2}\right]+2A_{2}C$$

Here,

H: $(4\pi^2 n^2 (dn/dC)^2/(N_A \lambda^4))$ optical constant;
C: concentration;
R_{vv}: Rayleigh ratio;
M_{w}: weight-average molecular weight;
R_{g}: radius of gyration;
q: $(4\pi/\lambda)\sin(\theta/2)$, scattering vector;
A_2: second virial coefficient.

The subscript vv denotes vertically polarized incident and scattered light.

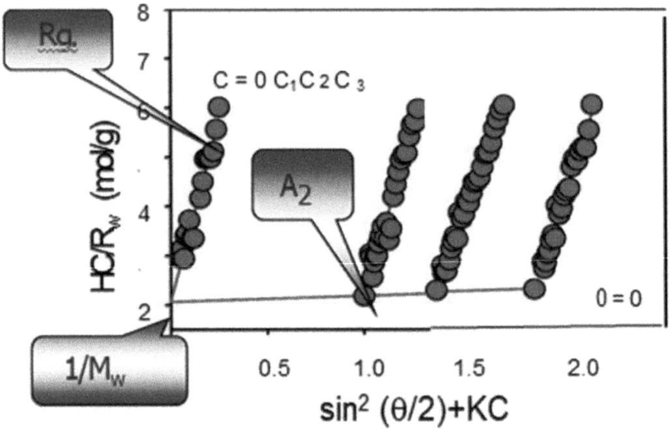

Figure 2.21 Zimm plot based on Section 2.5.4. Note: C = 0, C_1, C_2, C_3. By extrapolating the scattering angle (θ) to zero, we get A_2. By extrapolating the concentration (C) to zero, we get $1/M_w$.

2.5.5 Dynamic Light Scattering

The key equation for DLS is shown as follows:

$$\frac{\Gamma}{q^2} = D(1 + k_d C)[1 + f(R_g, \eta)^2]$$

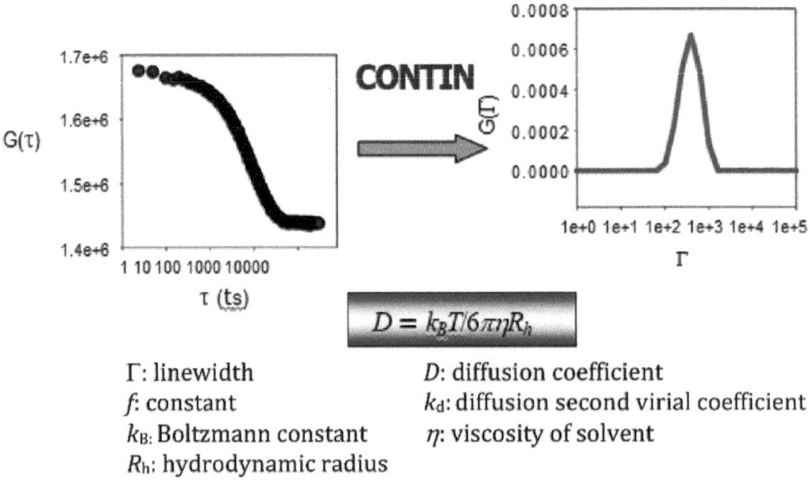

Figure 2.22 Dynamic light scattering based on Section 2.5.5.

We state that the average line width, <Γ>, is connected with the diffusion coefficient, D, by combination of a concentration term and a size-dependent correction term. The $G(\tau)$ and $G(\Gamma)$ terms are connected via the CONTIN procedure, a general-purpose constrained regularization program, primarily developed by S. W. Provencher, for analysis of DLS experiments from polymer solutions.

2.5.6 Laser Light Scattering (Example)

We shall use an actual laser light scattering experiment, including both static light scattering (SLS) and DLS, as an example, to demonstrate the type of information available from such studies.

	M_w (10^6 g/mol)	A_2 (Mol cm³ g⁻²)	R_g (nm)	R_h (nm)	D (10^{-8} cm² s⁻¹)	R_g/R_h
pCMVb plasmid 7164 bp in 1xTE	4.81 ± 0.21	8 × 10⁻⁴	135 ± 6	87 ± 2	2.78 ± 0.13	1.6 ± 0.1
94% DMF +6% water	6.9 ± 0.6	-3.7 × 10⁻⁴	15 ± 2	17 ± 2	12.4 ± 1.1	0.88 ± 0.15

Burchard, W. *Adv. Polymer Sci.* 1983, **48**, 1-123

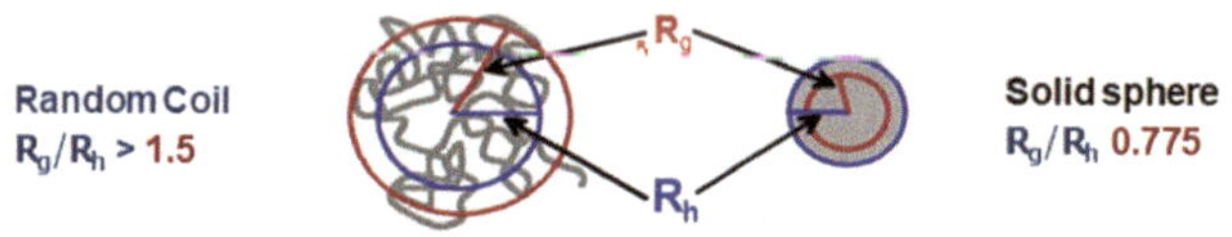

Figure 2.23 Information based on Laser Light Scattering.

Based on the discussions of laser light scattering experiments of macromolecular solutions, the key information that can be obtained is being presented by using an actual DNA solution in 1xTE buffer and a DMF and water solvent mixture, as shown in the above figure. In essence, the study can yield information on the molecular weight (M), the second virial coefficient (A_2), the radius of gyration (R_g), the hydrodynamic radius (R_h), and the diffusion coefficient (D). An interesting finding is the values of R_g/R_h ratio. By varying the solvent quality, we can obtain changes in the conformation of the macromolecule from a random coil to a "solid" sphere as has been demonstrated by the results indicated for a coil as shown in the upper row for the grey-scale figure (or yellow row in color) and for a "solid" sphere (lower row with darker shade for the grey-scale figure) (or magenta row in color). A schematic diagram below the table shows how the R_g/R_h ratio changes from a random coil to a (compact) "solid" sphere. Thus, even with a straightforward laser light scattering study, we can learn a great deal on how macromolecules behave in solvents of varying quality.

2.6 Synchrotron X-Rays

Now, we have gone through introductory laser light scattering with a single incident laser beam. Let us take a breather and switch to X-rays to remind us of the Interconnections between laser light and synchrotron X-rays.

It is interesting to realize that due to the brilliance of synchrotron light sources, the time for making measurements in an experiment is greatly being shortened. This shortened experimental time opens up new frontiers for time-dependent measurements. At the same time, new detectors and data storage schemes have been developed. Under such an environment, associated instruments trying to take advantage of synchrotron facilities must also consider the space available for new equipment and the integration of systems as well as compatibility at different synchrotron light sources, in order to be able to carry out experiments over limited time periods at a synchrotron facility. Furthermore, the additional supplementary instrumentation should be designed in such a way so that those instruments can also be used in a laboratory setting in order to maximize their utility and to justify their cost.

Let us first consider the changes in data accumulation and experimental project duration.

2.6.1 Experimental Project Duration

The key characteristics for example synchrotrons being considered are: 2nd generation: 2.8 Gev synchrotrons, e.g., NSLS, versus 3rd generation: 6.8 GeV synchrotrons, e.g., ESRF, APS, and Spring-8.

The ability to obtain experimental measurements over very short time intervals has been discussed in Section 1.2.1 in Chapter 1, "Synchrotron X-rays," where the capability to obtain information on time-dependent changes of state of systems has been discussed. At the same time, the requirements on instrumentation demand the need for expensive detectors and faster data analysis capabilities are briefly presented.

2.6.2 Key Factors for Efficient Use of Modern Synchrotron X-Ray Sources

Synchrotron X-ray sources are expensive instruments. It is essential to be able to collect a minimum set of data in order to be able to answer specific pertinent questions. Thus, the experimental plan implies that we need to consider on the

- use of best detectors available, although this step is *Not* associated with the major expense in the overall experimental cost;
- need for complementary instrumentation to analyze large amounts of experimental data.

Thus, we should pay even closer attention to sample preparations. An example for a synchrotron X-ray experiment may include data generation by using an area detector in the time-resolved mode. For example, with 10^6 pixels per frame at 10 sec per frame, we get 3.6×10^8 data points per hour. Consequently, such an experiment demands that we should pay close attention to sample preparation, i.e., we need to consider the cost associated with measurements for each sample.

2.6.3 Pinhole Collimation System [11]

We consider a simple pinhole system in order to provide the reader some basic information on the experimental needs for a three pinhole small-angle X-ray scattering collimation system:

- Pinhole on *tantalum* substrate with *tapered* edges – aiming to produce a well-defined X-ray beam;
- Multiple pinhole dimensions for optimization of d-spacing resolution versus flux;

- Pinholes, instead of slits, in order to achieve relatively easier alignment;
- With first pinhole of 0.1 mm diameter, a d-spacing resolution of 100 nm has been achieved.

2.6.4 SWAXS Setup and Sample Chambers

Information on the detectors used is to provide some idea on the scattering angular range that has been achieved.

2.6.4.1 Detectors used[a]

- 1-D SWAXS
- 2 Braun PSDs using PC for direct memory access
- 2 EMBL PSDs with CAMAC histogram memory module (64 MB)
- 2-D SWAXS
- Image plates for SWAXS measurements during still or steady state conditions
- EMBL 2-D MWPC for dynamic SAXS and WAXD measurements

2.6.5 Additional Instrumentation with Different Forms of Sample Chambers

Before we start with the additional instrumentation, the reader is reminded of Section 1.2.2 in Chapter 1, "Supplementary Instrumentation for Synchrotron X-rays," where a range of supplementary instruments have already been presented for applications to synchrotron beam line facilities. Here we have added a set of four additional instruments that have been adapted for use in combination with a synchrotron beam line facility. It should also be worthwhile to again remind the reader of the dual capability of using such instruments in a laboratory environment. Furthermore, if the probing radiation is an electromagnetic radiation in the visible to infrared range, laser light scattering is a laboratory instrument with different versions available commercially.

---

[a]Undoubtedly, new detectors will continue to become available.

We should also remember that instruments used for *perturbing* the sample can also be combined with instruments designed to *measure* physical parameters of the sample of interest. Thus, space requirements should be considered in advance when trying to integrate a range of instruments in order to avoid costly modifications.

In the figure on additional instrumentation (Figure 2.25), there are "four" units (if we count the two rheometric instruments as one unit): a draw unit, a stretch unit, a shear unit, and two rheometric instruments, that have been integrated with the synchrotron X-ray beam line for SAXS/WAXD measurements.

The draw unit is used to study the structure of fibers under extension. The figure on the draw unit with an X-ray collimator, as shown in the upper left panel on additional instrumentation has letter characters that may be too small for comfortable visual observation. Thus, we have repeated the explanation as shown below (*Clockwise from 12 o'clock*):

Hysteresis Brake, Roller Guide, Head Pins, Yarn Uptake (located at 6 o'clock and pointed by an arrow), 2 Godet Rolls, Separator Rolls, Yarn Supply (Yarn in a different shade color). Synchrotron X-rays (denoted by a thick line) passing through the Pinhole Collimator is located below the Head Pins with the 2D X-ray scattering/diffraction pattern being illustrated by the darker blue panel. In the colored presentation those identifications become clearer.

In the picture on the left of the draw unit, the large dark colored unit is the draw unit and the cream colored unit is the Mar 2D X-ray detector.

The remaining three units are designed to investigate the structure or rheological properties of materials at different temperatures when the sample is being stretched or sheared.

For the synchrotron SAXS/WAXD beam line, it is also feasible to combine X-ray detection with other forms of physical measurements, such as Raman scattering, being shown schematically in Figure 2.24.

The key point is that optical fibers can be used for a variety of applications. Here we have combined an optical Raman probe in combination with the SAXS/WAXD beam line. The supplementary instrumentation being mentioned represents examples whereby we can perturb the sample of interest while providing SAXS/WAXD measurements in order to observe the structural changes while the Raman probe is to provide additional information on the sample of interest by combining the X-ray measurements with Raman spectroscopy. There are undoubtedly other combinations on either perturbing the sample of interest or introducing other forms of detection on sample structure changes. The two figures which we have presented are representative examples for the reader to be aware of such combinations. It is worthwhile to design a synchrotron beam line facility with some space flexibility for future accommodations on supplementary instrumentation.

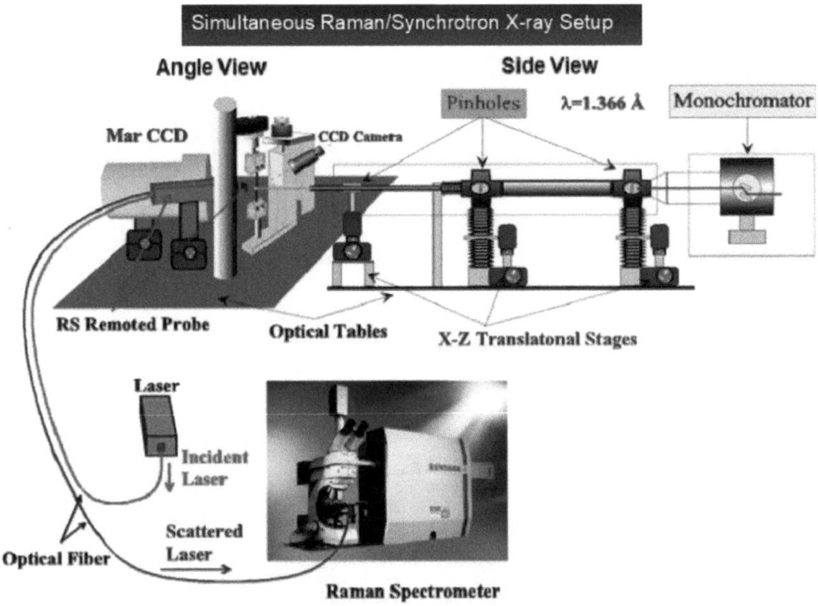

Figure 2.24 Simultaneous Raman-synchrotron X-ray setup.

It should be noted that the synchrotron X-ray facility operates on a different scale and schedule when compared with laboratory instrumentation. Adaptation and appropriate adjustment for taking advantages of such large facilities is a worthy effort that the user should understand.

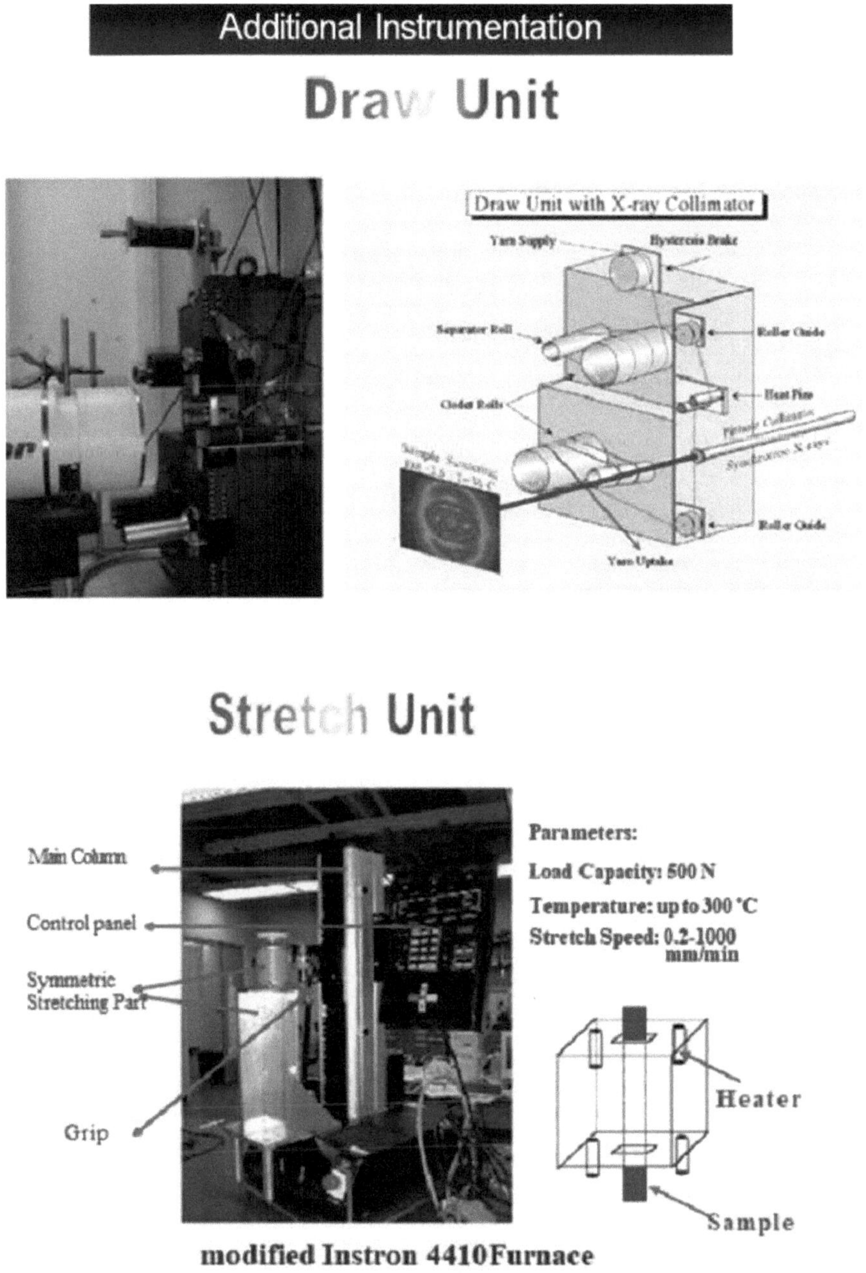

Figure 2.25 Draw and stretch unit.

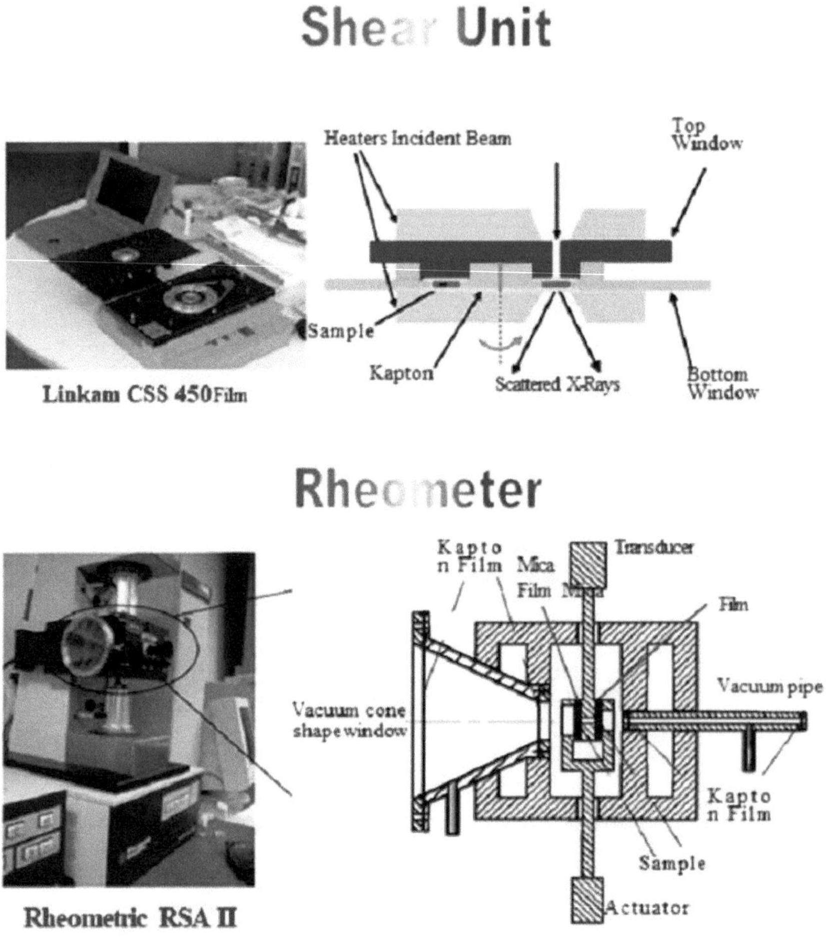

Figure 2.26 Rheometer and shear unit.

2.6.6 Research Opportunities via Synchrotron X-Ray Scattering/Diffraction

2.6.6.1 Example: SAXS/WAXD at X27C for advanced materials

Examples for advanced materials include their changes in supra-molecular structures of biological/biosynthetic macromolecules, ionomers, block copolymers, dendrimers, charged gels and colloidal systems, e.g., nanoparticles of organic/inorganic materials and with different shapes, in solutions, suspensions, and melts. The studies cover a broad range, such as

- Kinetics of Phase Transition in Semi-crystalline Polymers (Crystallization and Melting), Multicomponent Mixtures and Molecularly Segmented Systems (Phase Separation);
- High Pressure Experiments (Hydrostatic Pressure, Super-critical Solvents) for Environmental Process Development;
- *In-situ* Chemical Reaction and Polymerization;
- Polymer Fibers/Films during Processing (Spinning, Extrusion, Deformation) Polymer Melts/Solutions in Flow.

2.6.6.2 Dynamic light scattering optical schemes

Returning to laser light scattering, we shall now start with photon cross-correlation by first considering DLS optical schemes and 3-dimensional dynamic light scattering (3D-DLS), whereby the X-ray instrumentation shall be designed to have capability to couple with 3D-DLS, yielding simultaneous laser cross-correlation and synchrotron small-angle X-ray scattering/wide angle X-ray diffraction (SAXS/WAXD).

In the optical schemes for dynamic light scattering (DLS-1), as shown in Figure 2.27, there are essentially three schemes: (a) a one-beam light scattering optical scheme which is being followed by most light scattering instruments; (b) Phillies scheme which is being included for completeness; and (c) a *two-color* dynamic light scattering scheme which deals with two crossed incident laser beams of current interest in our studies. The second vector diagram for 3D-DLS shows how the two incident laser beams ($\mathbf{k}_{i,1}$ and $\mathbf{k}_{i,2}$) cross and the scattering angle θ, the magnitude of momentum transfer vector ($\mathbf{q}$), etc., are determined.

Now, BC calls to the attention of readers on the complexities associated with two-color dynamic light scattering schemes, starting with Figure 2.28. It becomes even more challenging when such an instrument is connected with synchrotron X-ray experiments, as shown in Figure 2.32. Although the validity of such a combined laser light and synchrotron X-ray instrument has been demonstrated in Chapter 2, the last sections (Figures 2.27 to 2.40 and Figures 2.44 to 2.51) are intended only for experienced scientists/engineers. For most students, the section should be deferred until there is sufficient experience on standard light scattering and synchrotron X-ray setups.

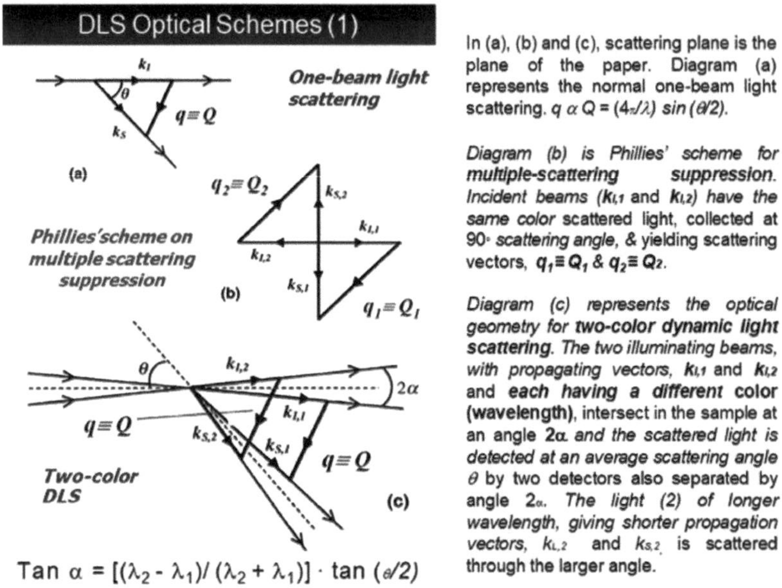

Figure 2.27 DLS Optical Schemes.

In diagrams (a), (b) and (c), of Figure 2.27 the scattering plane is the plane of the paper. Diagram (a) represents the normal one-beam light scattering. $q = Q = (4\pi/\lambda)\ sin\,(\theta/2)$. Diagram (b) is Phillies' scheme[b] for *multiple-scattering suppression. Incident beams* ($\mathbf{k}_{i,1}$ and $\mathbf{k}_{i,2}$) *have the same color* scattered light, at 90° *scattering angle,* and yield scattering vectors, $\mathbf{q}_1 = \mathbf{Q}_1$ and $\mathbf{q}_2 = \mathbf{Q}_2$. *Diagram (c) represents the optical geometry for* ***two-color dynamic light scattering****. The two illuminating beams, with propagating vectors,* $\mathbf{k}_{i,1}$ and $\mathbf{k}_{i,2}$, and ***each having a different*** **color (wavelength)**, intersect in the sample at an angle 2θ *and the scattered light is detected at an average scattering angle* θ by two detectors also separated by angle 2θ. *The light (2) of longer wavelength, giving shorter propagation vectors,* $\mathbf{k}_{i,2}$ and $\mathbf{k}_{S,2}$, is scattered through the larger angle, as shown in Figure 2.28.

The geometrical figure, as shown in Figure 2.28, represents a complex three dimensional (3D) concept for 3D dynamic light

[b]George Phillies, Professor of Physics, Emeritus, Worcester Polytechnic Institute, Worcester, MA.

scattering (DLS) which may become a challenge for the less initiated reader to visualize. Thus, it is advisable to read over the descriptions below very carefully. It is also normal for beginners to read over the description, together with Figure 2.28, more than once and pay attention to the different orientations being shown above and below the orientation of the scattering plane (AB).

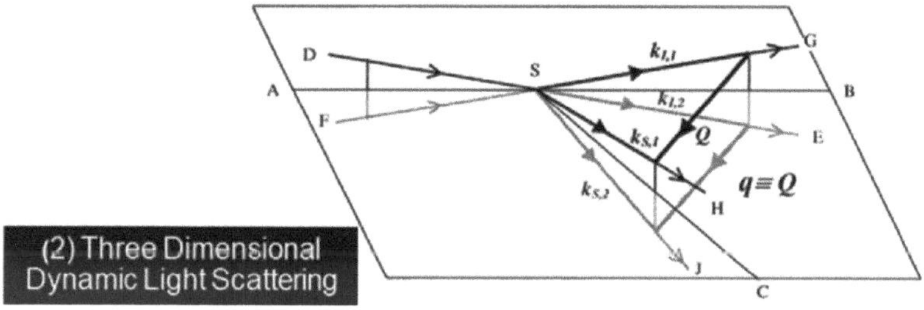

Figure 2.28 Three-Dimensional dynamic light scattering.

The parallelogram in Figure 2.28 represents the *average* scattering plane. **Solid** lines are **above** the scattering plane, **shaded** (grey) lines **below** the scattering plane. Illuminating beams, DE and FG, originating from above (for both DS and SG) and below (for both FS and SE) the average scattering plane, crossing at point S, and propagating below (for SE) and above (for SG) the average scattering plane, have the propagation vectors, $\mathbf{k}_{i,2}$ and $\mathbf{k}_{i,1}$ respectively. AB represents the perpendicular projection of the two beams, DE and FG, onto the plane. The scattered beams, SH (above the plane) and SJ (below the plane) project onto SC. The scattering angle θ is the angle between SG and SH and between SE and SJ. The beam pairs, SG-SH and SE-SJ, define the same scattering vector $\mathbf{q}$ (or $\mathbf{Q}$).

2.6.7 Overview of Simultaneous Laser Cross-Correlation and Simultaneous SAXS/WAXD

A schematic diagram for 3D-DLS is shown in Figure 2.29. Here, we have two incident laser beams coming from lasers (LD1 and LD2) with two photo-detectors (PD1 and PD2) for the scattered light. Figure 2.29 shows both the side view and the top view of the setup for 3-dimensional dynamic light scattering.

The setup for 3D-DLS, as shown in Figure 2.29, uses two diode lasers (LD1 and LD2) impinging on sample (S) at the crossover point, with the coordinate system being shown in the inset of Figure 2.30 (3D-DLS, side view). Pay attention to the location of G, D, E, and F in Figure 2.30. 3D-DLS can also be achieved by using a beam splitter with a single laser as shown in Figure 2.31 for 3-dimensional DLS.

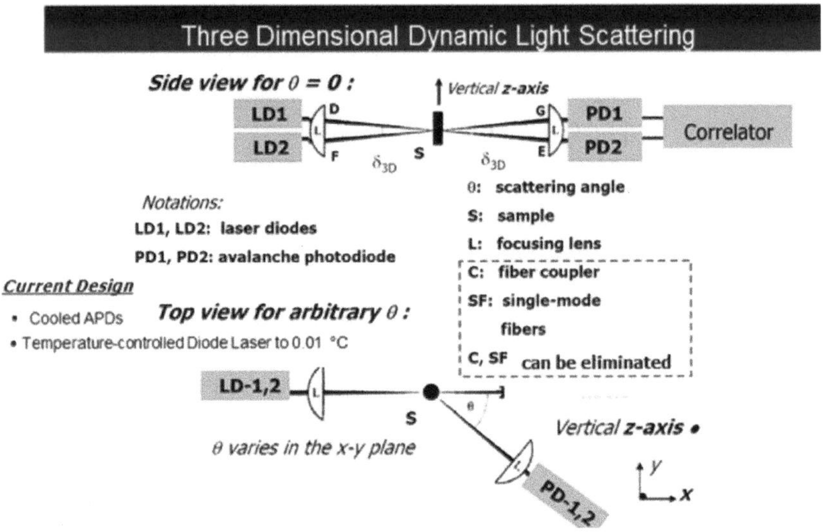

Figure 2.29 Three Dimensional Dynamic Light Scattering.

Figure 2.31 on 3D-DLS shows a different incident laser beam setup whereby the incident laser has been split up to have two incident laser beams (beam 1 and beam 2) and the symmetry in design of the setup for 3D-DLS. The two Figures (2.30 and 2.31) show mainly the difference of having to use two diode lasers (LD1 and LD2) or a single laser with a beam splitter and mirror to produce the crossing of two incident laser beams in the setup, as shown in 3-dimensional DLS. However, the key point is the Figure (2.31) on optical geometry and side view of 3D-DLS where we have vectors $\mathbf{k}_{i1}$ and $\mathbf{k}_{i2}$ for the two incident beams, and $\mathbf{k}_{s1}$ and $\mathbf{k}_{s2}$ for the two scattered beams, with the shaded scattering area, as shown in the upper left figure, defining the magnitude of the momentum transfer vector ($\mathbf{q}$) and the scattering angle (θ). The angle of crossing of the two incident laser beams (δ, as shown in Figure 2.31), is shown in the side view of 3D-DLS in Figure 2.30.

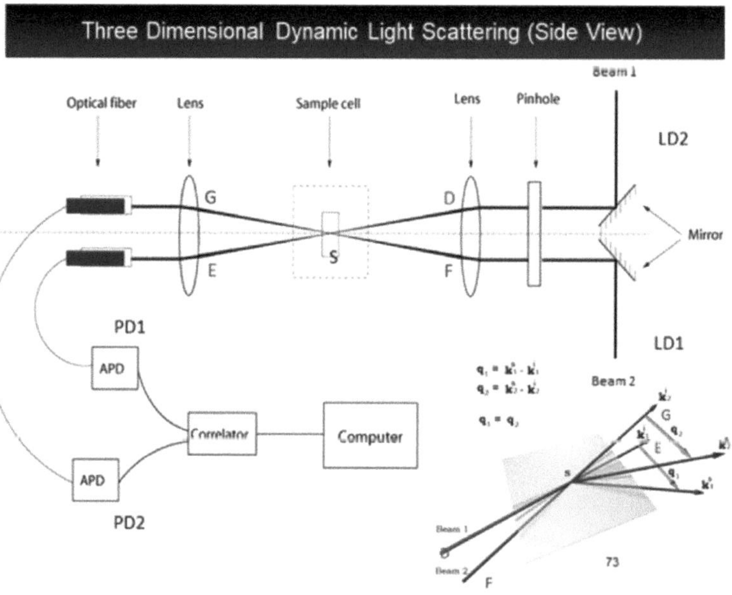

Figure 2.30

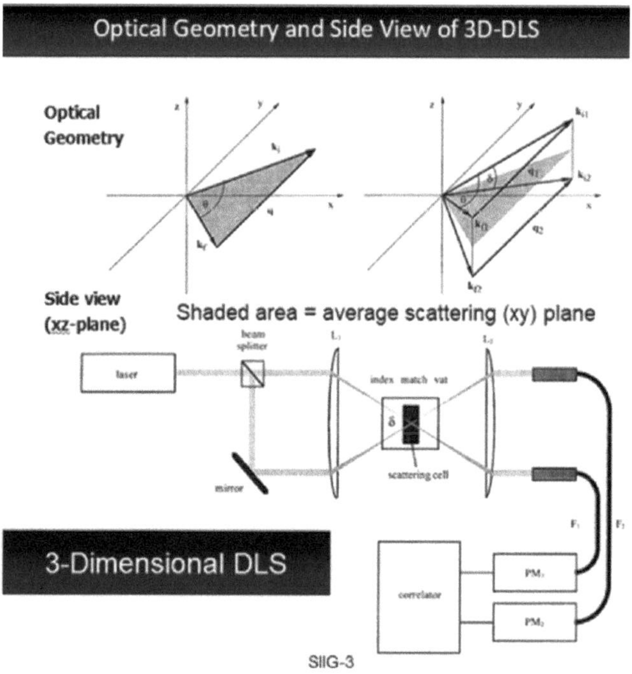

Figure 2.31 Optical Geometry and Side View of 3D-DLS.

It is important to pay attention to the coordinate system being shown on the lower right side of Figure 2.30 where G, E, D, and F have been shade coded, with S denoting the crossing point of the two laser beams crossing with each other. S also denotes the location of the sample of interest in the Figures, 2.29 and 2.30.

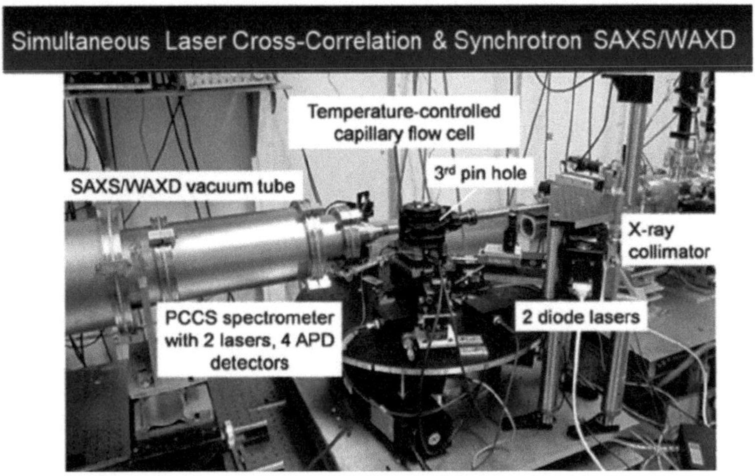

Figure 2.32 Simultaneous laser cross-correlation and synchrotron SAXS/WAXD. Automated rotary table with translational stages is used for cell centering and movements of the sample chamber into the synchrotron X-ray beam. The combination of light/X-ray scattering setup is on the same optical table as the synchrotron X-ray beam collimation system to establish instrument stability.

To combine 3D-DLS with synchrotron X-rays, we show two figures:

(1) Side, top, and enlarged top views of laser beam pathways in Figure 2.33, and
(2) Side and top views of X-ray pathways, as shown in Figure 2.34.

It should be noted that in the simultaneous laser cross-correlation and synchrotron SAXS/WAXS, as shown in Figures 2.35 to 2.40, we should pay attention to how the various components have been integrated in the overall system. Further details have been presented on the schematics of laser cross-correlation instrument with drawings in combination with pictures.

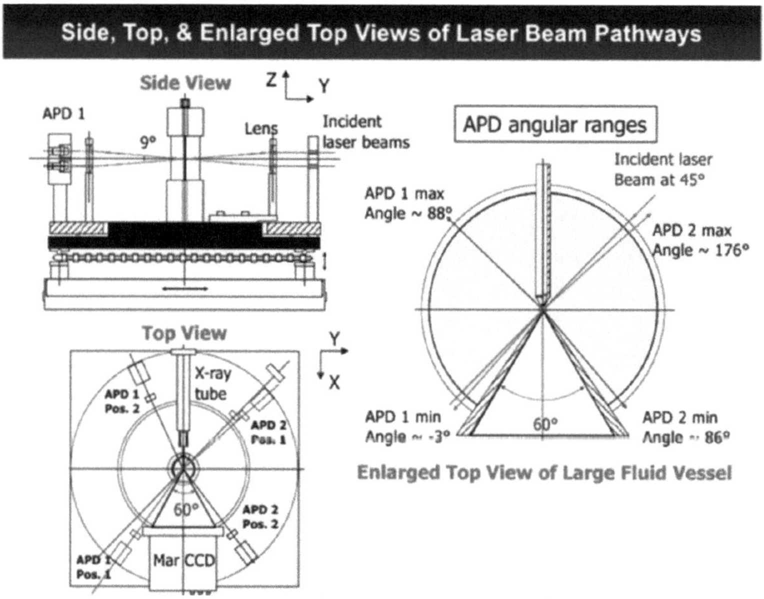

Figure 2.33 Side, top, and enlarged top views of laser beam pathways.

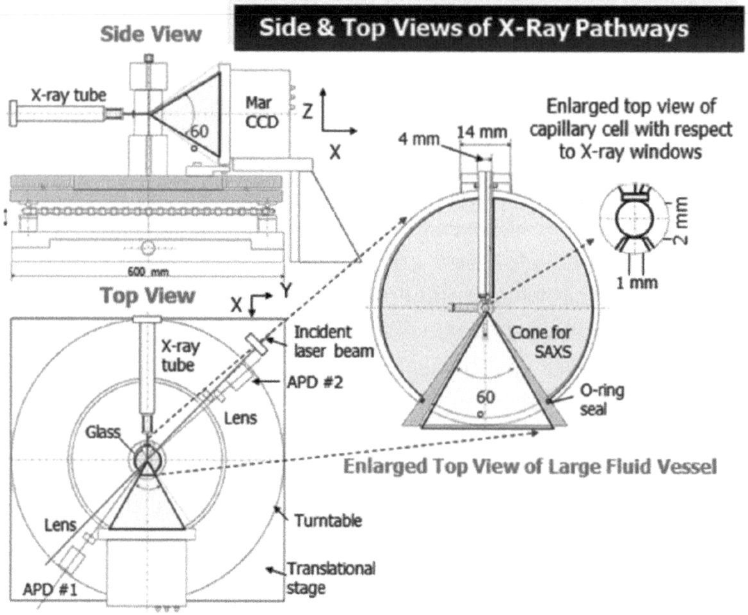

Figure 2.34 Side and top views of X-ray pathways.

The 3D-DLS apparatus is also being attributed to Professor Dufei Fang, who participated in the instrumentation design. He has developed a combination of pictures and drawings showing the details of the instrument that has many space, optical, and mechanical restrictions. Thus, his collaboration with me on the development of this synchrotron beam line is being acknowledged and appreciated. The APD angular ranges are shown on the top right side in the side, top, and enlarged top views of laser beam pathways. For the two APDs, APD-1 shows a min angle and a max angle of ~3° and ~88°, while for APD-2, min angle and max angle are, respectively, ~86° and ~176°.

The cylindrical scattering cell of 2-mm in diameter has been used successfully and is shown schematically in Figure 2.34 on side and top views of laser beam pathways as an enlarged top view of the capillary cell with respect to the X-ray windows in the upper right corner of Figure 2.34.

Note the X, Y, Z coordinate system in the laser beam and X-ray pathways. Details of combined laser light scattering and synchrotron X-ray scattering are presented in Figures 2.33–2.40.

Figures 2.35–2.40 represent a combination of 3D-drawings and pictures to show some details on how the simultaneous laser cross-correlation and synchrotron SAXS/WAXD instrument was assembled. In each schematic drawing, the essential components are presented, leaving out the other parts. The reader may wish to view the schematic drawing with the associated photographs. The six figures are as follows.

- Schematics of Laser Cross-Correlation Instrument: To show the two diode lasers and the two APDs in relation with the flow cell where other associated components have been removed, as shown in Figure 2.35.
- Dual laser and focusing unit: To show the translational controls and components of the two diode lasers, as shown in Figure 2.36.
- X-ray scattering/cross-correlation DLS system: To show the two diode lasers in combination with the two APD detectors, as shown in Figure 2.37.
- Top View (section) of cell assembly by light and X-rays: An overview on how both laser light and synchrotron X-rays can reach the sample of interest. Pay attention to two earlier

drawings on laser beam and X-ray beam pathways, as shown in Figure 2.38.

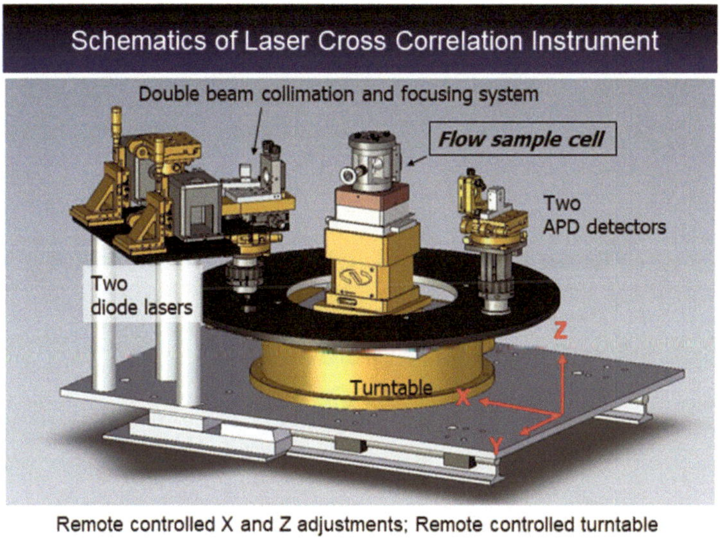

Figure 2.35 Schematics of laser cross-correlation instrument.

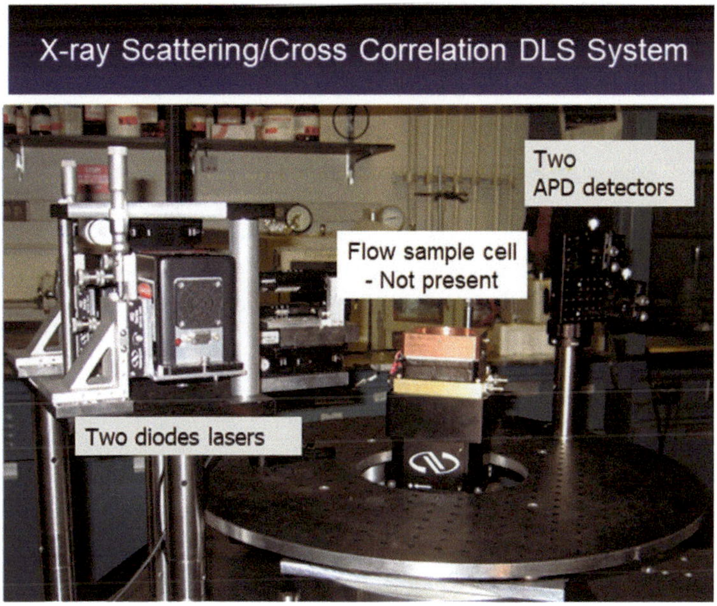

Figure 2.36 X-ray scattering and cross-correlation DLS system.

- Light scattering and X-ray scattering cell: To show a combination of drawings and pictures, describing the relationship between both light and X-rays can reach the sample and the scattering/diffraction information from both light and X-rays, as shown in Figure 2.39.
- Flow sample cell with quartz capillary: In the X-ray pathway figure, a capillary cell with a diameter down to 2 mm has been achieved, as shown in Figure 2.40.

Figure 2.37 Dual laser and focusing unit.

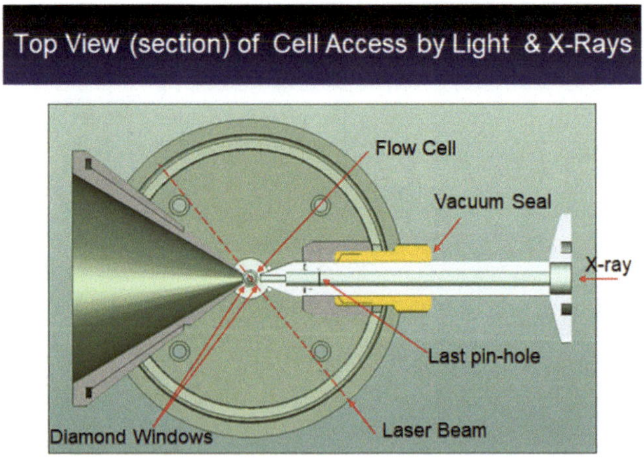

Figure 2.38 Top view of cell access by light and X-rays.

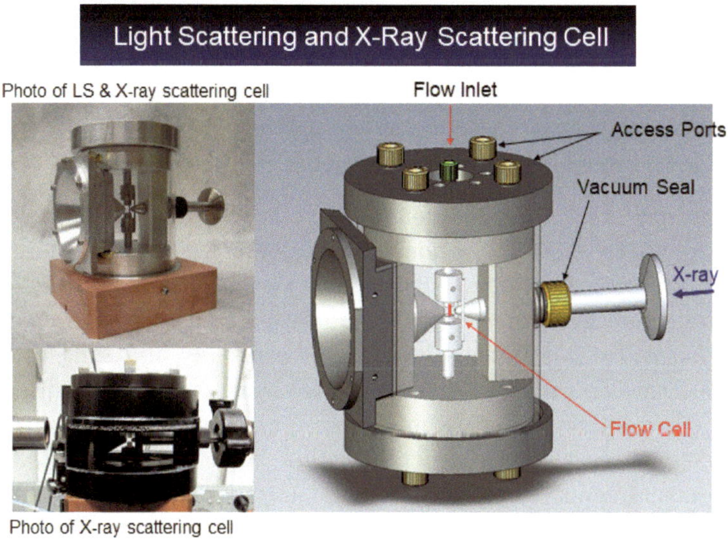

Figure 2.39 Combined laser light and X-ray scattering cell.

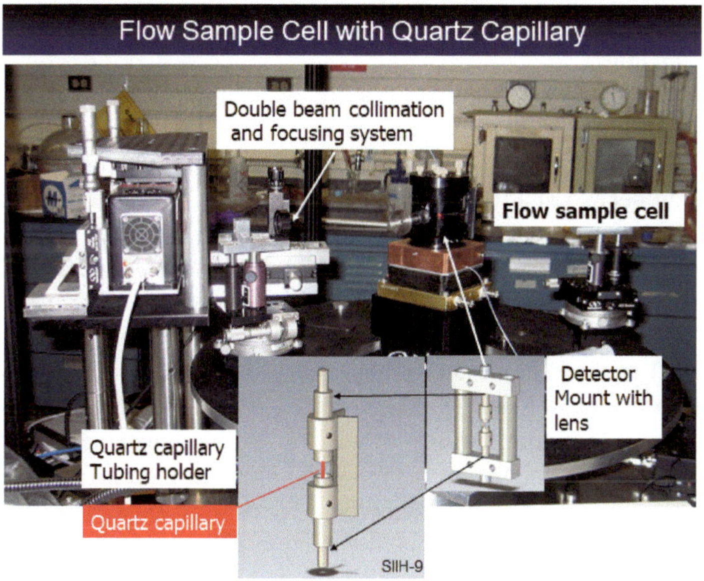

Figure 2.40 Quartz capillary flow sample cell for simultaneous laser light scattering and synchrotron X-ray scattering measurements.

In the photo of X-ray scattering cell, you may wish to look close at a part of the LS and X-ray scattering cell showing in the middle of the much larger black thermostat with the right side having the X-ray inlet and the left side a cone (also being shown in the photo of LS and X-ray scattering cell).

2.6.8 Designs on Overall Synchrotron Setup

We now use the experimental setup of the Advanced Polymers Beam Line at X27C, National Synchrotron Light Source, located at Brookhaven National Laboratory on Long Island, New York to demonstrate some of the desirable features in the design of a synchrotron X-ray beam line. As the capability of a synchrotron X-ray facility can be further broadened by incorporating additional instrumentation that permits measurements of complimentary physical parameters in combination with synchrotron X-ray measurements, more co-operative information becomes feasible. Thus, a few pertinent examples have been presented below. We further design our synchrotron X-ray scattering/diffraction facility with built-in features that permit easier introduction of other instruments with a stable platform already incorporated into the beam line.

The use of commercial readily available optical tables is one possible route, worthy of consideration. The platform permits (1) the development of a flexible, stable and well-defined incident X-ray beam, (2) a flexible sample chamber, (3) a detection system that can accommodate different types of detectors, and (4) attachments of other instrumentation to expand the capability of the synchrotron facility.

In Figure 2.41, the short smaller platform in the front (upper right) is the collimation system showing how the incident X-ray beam can be defined. The larger optical table after the collimation system shows the vacuum tube with flexible length features for scattering/diffraction measurements. For the figure, a Mar CCD detector is shown on the lower left.

It should be noted that the sample chamber is between the incident beam collimation system and the long vacuum tube together with the Mar CCD detector. The space between the incident beam collimator and the vacuum tube before the CCD detector is variable, as can be seen by the tracks on the optical tables.

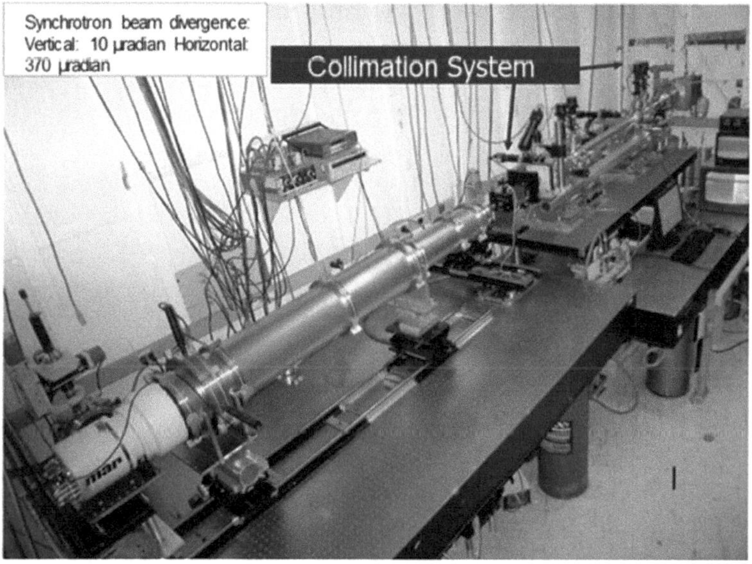

Figure 2.41 Advances Polymers Beam Line, X27C, National Synchrotron Light Source, Brookhaven National Laboratory, LI, NY.

To define the incident synchrotron X-ray beam, we show the 3 pinhole geometry (Figure 2.42), with d_1 and d_2 defining the incident X-ray beam from the X-ray source and the 3rd pinhole, d_3, acting as the guard pinhole.

It should be noted that a set of pinholes with different pinhole diameters are prepared for the collimation system, depending on the experimental needs. More pinholes can be fabricated to accommodate the beam line for additional demands. Fixed diameter pinholes are well defined and can be checked for future references.

The synchrotron beam has its own beam characteristics. The pinhole (or slit) system is used to further refine the incident synchrotron X-ray beam.

An overall side view picture of the pinhole X-ray collimation system is shown in Figure 2.43. It should be noted that the height from the optical table base was mainly controlled by stacking of materials so that the critical distance from the base to the synchrotron X-ray beam could be dependent mainly by the thermal coefficient of expansion of the materials being used and the controlled room temperature in the hutch, housing the beam line.

The four (black colored) vertical micrometers, with two in the front and two in the back, are for the movements to control the alignment of the two pinholes that define the incident synchrotron X-ray beam. The translational motions for the two pinholes were being controlled to the order of microns. Again, in order to provide the pinhole stability of the incident X-ray collimator, it is worthwhile to take into account the thermal stability of the collimator from the ground up, implying the control of the thermal coefficient of expansion of the collimator from the ground up and temperature control of the instrumentation chamber.

It is important to realize that the mechanical stability of the beam line depends on its relationship with the synchrotron beam coming from the "source". Thus, its physical stability is coupled with the stability of the instrument from the base (ground). We have paid attention to the thermal coefficient of expansion of the materials used in constructing the beam line from ground up.

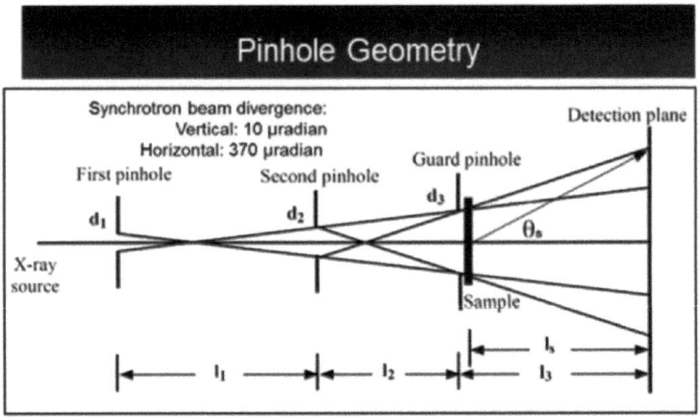

Smallest angle without serious parasitic scattering, θ_s:

d_s denotes a finite size for the detector element, and d_3 [$= (d_1 + d_2) \bullet (l_2/l_1) + d_2$] depends on l_1, l_2, d_1, and d_2

Key points:

1. Minimization of required translational stages
2. Succeeded in reaching theoretical limit
3. Discrete variation in incident beam cross section

Figure 2.42 X-ray pinhole geometry.

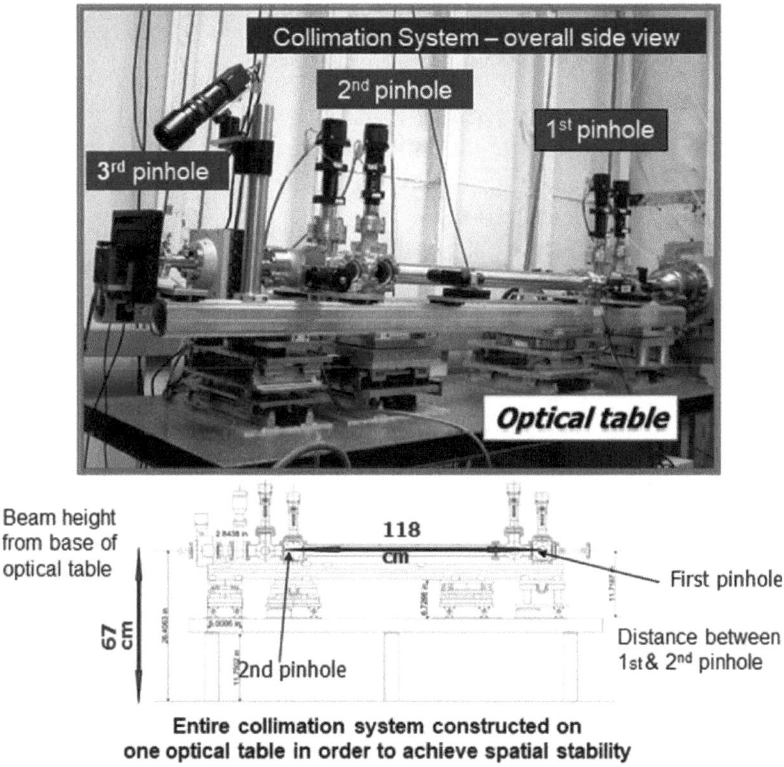

Figure 2.43 Collimation system with overall side view and instrumentation dimensions.

The optical table base, as shown for the collimation system in the picture below, sits on a larger optical table, as shown in the Advanced Polymer Beam Line, X27C which no longer exists.

2.6.9 Test on Instrumentation Stability (by Using the Synchrotron Beam Line at X-27C)

Aqueous suspensions of (mono-disperse) 50 nm diameter polystyrene (PS) beads were used to demonstrate the scattering behavior of such a beam line, as shown in Figure 2.44. Note the weak scattering of the PS latex suspension when compared with the solvent. After the solvent subtraction, the scattering of the polystyrene beads are shown in the curve below (after solvent subtraction, as shown in the upper right corner), with intensity being shown in log scale. Yet the signal to noise ratio is still acceptable.

It should be noted that we have intentionally selected polystyrene latex spheres as the reference dispersed material, instead of, for example, gold spheres. The weak scattering behavior tries to demonstrate the ability to obtain information even near the limit of solvent scattering.

Note that for Figure 2.44 below, the intensity in the inset is on a linear scale, while that for the figure is on a logarithm scale.

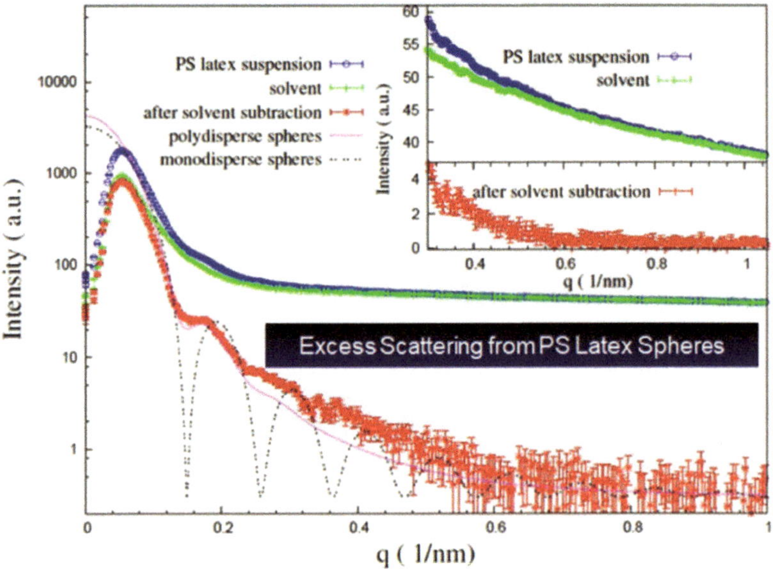

Figure 2.44 Aqueous suspension of polystyrene latex with sphere diameter (D = 50 nm).

Having had confidence on our pinhole collimation system for synchrotron X-rays and our earlier presentation of a single incident laser beam dynamic light scattering system, we now proceed to find out how we can combine laser light scattering, not only with a single incident laser beam but also with two incident crossed laser beams, essentially 3D-DLS, in combination with synchrotron X-ray scattering measurements.

In the simultaneous laser cross-correlation and synchrotron SAXS/WAXD experimental setup (Figure 2.45), the X-ray collimator has remained the same as we have discussed before. Similarly, we have used the two diode incident laser beam configuration with the schematics that has already been presented. The sample chamber space has been enlarged to accommodate a rotating table (to be described soon afterward). The compact photon cross-correlation spectroscopy (PCCS) spectrometer is integrated with the synchrotron X-ray instrumentation, whereby the 3rd X-ray pinhole is located inside the temperature-controlled capillary flow cell, also to be presented in later figures.

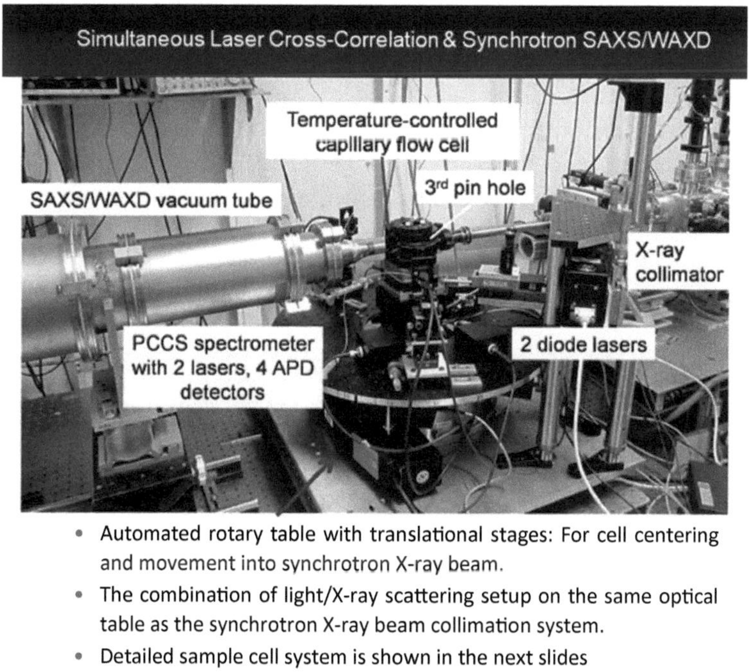

Figure 2.45 Simultaneous laser cross-correlation and synchrotron SAXS/WAXD. Details of combined laser light scattering and synchrotron X-ray scattering are presented in Figures 2.33 to 2.40.

In order to show how laser light beam and X-ray beam pathways can be accommodated with the integrated flow cell, we first show the laser beam pathways and the synchrotron X-ray beam pathways in two separate figures. It should be noted that when presenting those pathways, only the essential elements of the instruments associated with those beam defining elements are presented. All the components have their physical dimensions, available range limitation, and precision specifications that should be implicitly taken into account. For example, in the laser light beam pathways, we need to take into account the specific APD being used to determine the angular ranges and orientation of the instrument. The same consideration has been applied to the X-ray beam pathway where the third pinhole (under vacuum) needs to be delivered close to the sample cell of interest.

To test the validity of the instrument performance, we used an aqueous suspension of uniform polystyrene latex spheres with a diameter of 50 nm and at a concentration (c) of 5×10^{-4} g/ml. The scattering measurements were carried out at $\theta = 40°$ by using a 5 mm diameter cylindrical cell. The test is being used to confirm the validity of the experimental setup for cross laser beam experiments. By using a single incident laser beam or by using two crossed incident laser beams, the results agreed with the theoretical curves (dotted curves) for the polystyrene latex sphere suspension, as shown in Figure 2.46.

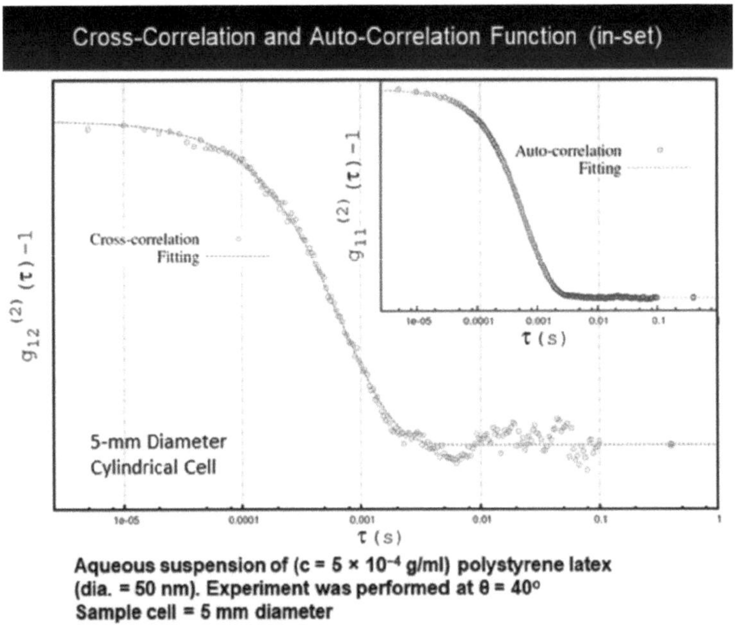

Figure 2.46 5 mm diameter quartz capillary flow sample cell for laser light scattering using both cross- and auto-correlation function measurements.

More detailed measurements of the auto-correlation function from the two separate incident laser beams with each laser beam being used independently as a single incident laser beam are shown in the auto-correlation function Figure 2.47, as Auto 1 and Auto 2. It should be noted that the sample concentration has been reduced by a factor of 10 to c = 5 × 10^{-5} g/ml for the polystyrene latex suspension. The cylindrical sample cell diameter has remained at 5 mm and the scattering angle (θ) = 40°.

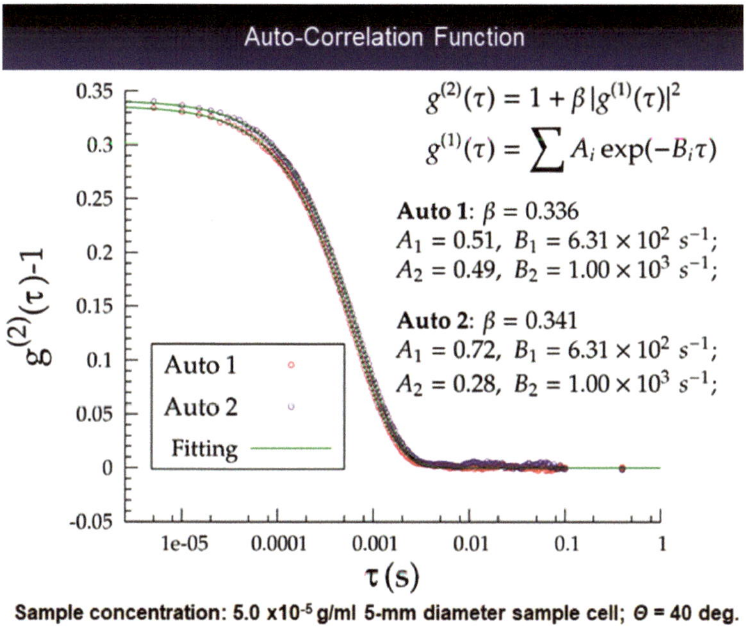

Figure 2.47 5 mm diameter quartz capillary flow sample cell for auto-correlation function measurements.

The cross-correlation function measurement was then carried out under the same conditions and agreed with the theoretical curve, as shown in Figure 2.48. It should be noted that the magnitude of the signal, as shown in the y-axis, is now substantially lower.

We also double checked further and used a 2-mm diameter capillary cell. The results are shown in the next two figures (2.49 and 2.50). For the cross-correlation function measurements, we increased the measurement time by a factor of 5 to 10 minutes.

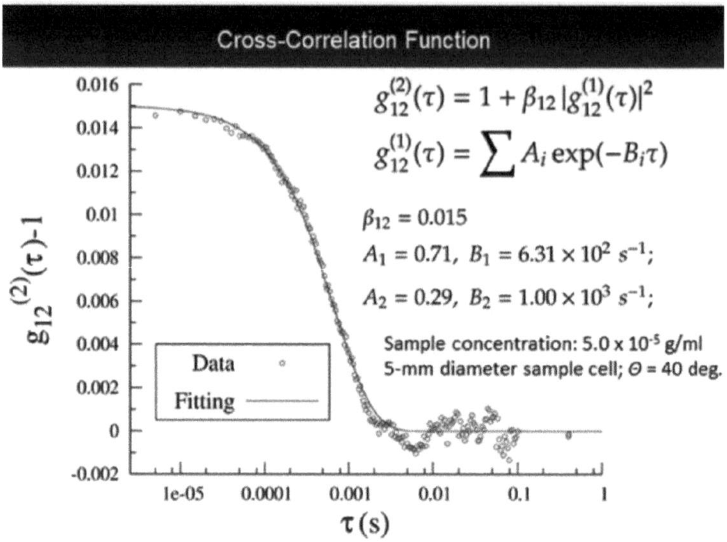

Figure 2.48 5 mm diameter quartz capillary flow sample cell for further cross-correlation function tests.

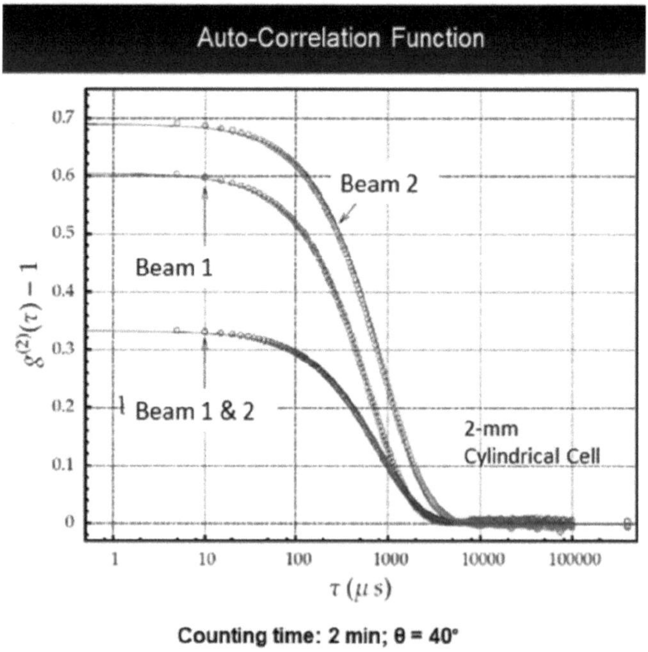

Figure 2.49 Further tests with a 2-mm diameter cylindrical cell.

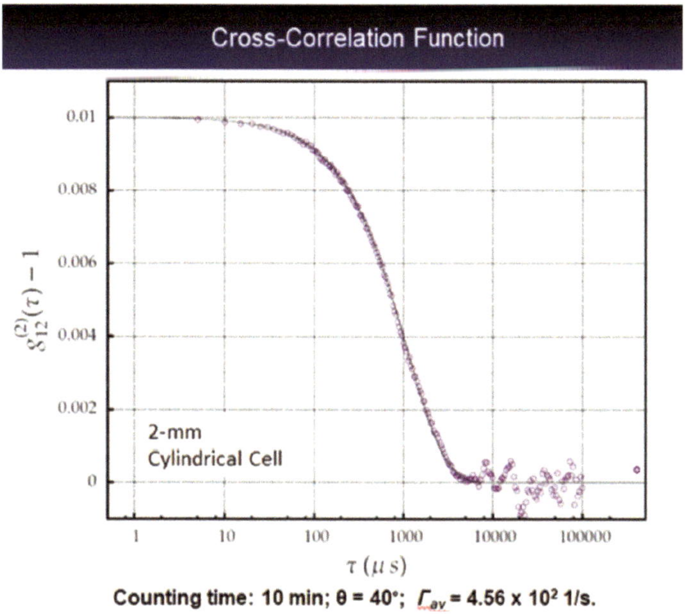

Figure 2.50 Cross-correlation function with a 2 mm cylindrical cell.

Results of the cross- and auto-correlation functions were compared further by using $G(\Gamma)$ versus Γ for cross- and auto-correlation functions, as shown in Figure 2.51.

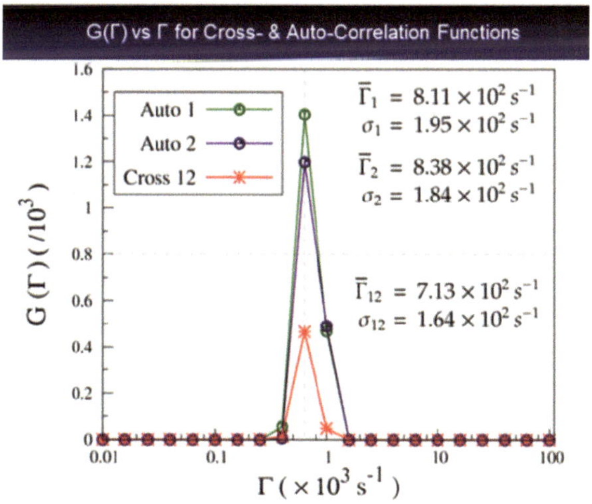

Figure 2.51

2.6.10 Partial Motivation for the Design and Construction of Laser Light/Synchrotron Small-Angle X-Ray Scattering in This Chapter

The figure on the structure, formation and inhibition of Amyloid fibrils has postulated early events in fibril formation that has emerged from atomic force microscopy (AFM) studies. The pathway on the Ab42 monomer, dimer, and tetramer formation will be of interest. The details will be presented in Section 4-G (SIV-G). The instruments being presented in this chapter are suitable for such in-situ studies. The topic is of importance because we have created a new tool to investigate Alzheimer's disease, Parkinson's disease, and prion disease.

The development of a combined laser light/SAXS-WAXD instrument, as has been described and tested in this chapter, is aimed at kinetic studies on 'aggregation' phenomena whereby a limited number of multiple species may undergo changes, such as size, shape, or/and aggregations.

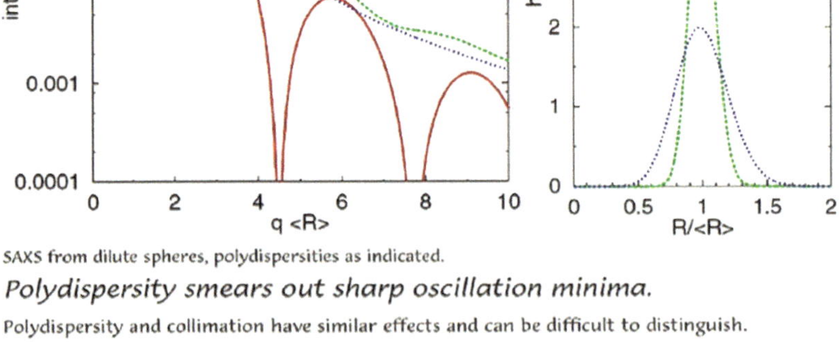

Figure 2.52 Intensity versus *q*<*R*> & *h*(*R*) versus *R* <*R*>.

The nature of scattering curves changes its shape when the particle changes its shape. In our calibration, we have used "mono-disperse" polystyrene latex spheres to demonstrate the behavior of auto- and cross-correlation functions. In practice, it is useful to have some idea on the scattering behavior of particles that have a size distribution, or non-spherical particles, which are, e.g., ellipsoids of revolution. Then, the scattering curves can be quite different, as shown in the following two figures. The intention for raising this point is again to remind the reader of the importance of boundary conditions associated with the instrument used, the sample under investigation, and the corresponding results.

It should be noted that X-rays scattered by electrons while light scattering depends on refractive index differences. Thus, some sense on how X-ray scattering is different from laser light scattering is another useful piece of information.

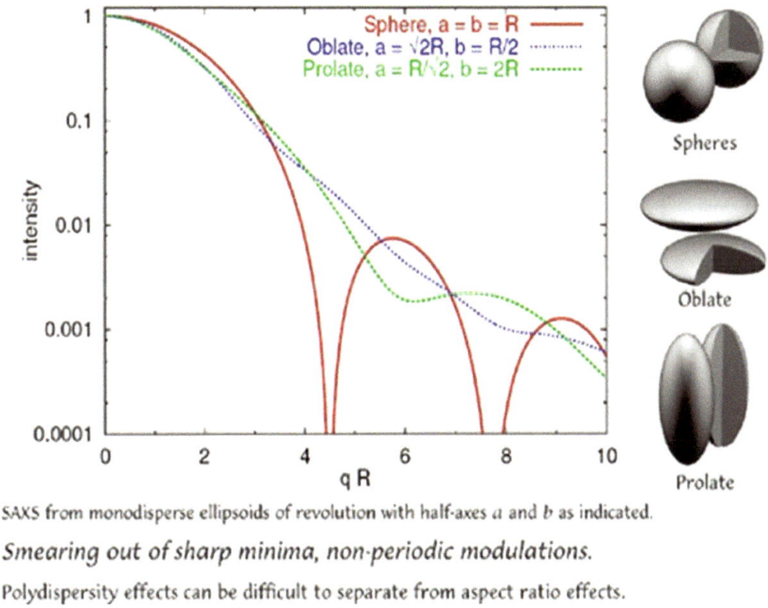

Figure 2.53 Comparison of Spheres vs. Ellipsoids of Revolution.

X-rays are scattered by electrons.

Scattering contrast between particle (or pore) and its solvent (or matrix) ~ ***square of electron density difference*** $\Delta\rho_{el}^2$.

Scattered intensity $I \sim \Delta\rho_{el}^2\, \phi\, (1 - \phi) \approx \Delta\rho_{el}^2\, \phi$ for small volume fractions ϕ.

Note that *all electrons* count for the electron density, not only valence electrons.
Heavy metals with lots of electrons have very high scattering powers.

substance	N_{el}/nm^3	rel.scat.power (water = 1)
N_2 (≈ air)	0.4	2×10^{-6}
water	268	1
graphite	650	5.9
diamond	1057	15.6
copper	2451	84
silver	2753	106
gold	4661	303

Figure 2.54 Scattering contrast of substances.

2.7 Take Home Message

This chapter provides an overview of the type of experiments that can be carried out by using laser light and synchrotron X-ray scattering for materials primarily related to macromolecular and supra-molecular systems over a very broad range of length scales. The take home message, being aimed at readers who are interested in utilizing these two techniques has three objectives:

(1) With emphasis on physics, Moiré interference patterns are being used to demonstrate light scattering spectroscopy in terms of homodyne, heterodyne, and self-beating techniques;

(2) Based on detection optics for one-beam and crossed two-beam light scattering, a more comprehensive description on laser light scattering is presented; and

(3) An overview of Simultaneous Laser Cross-Correlation and Synchrotron SAXS/WAXD is the final aim of this chapter.

We have emphasized the use of physics to explain the topics of current interest and the importance of having a solid foundation to build up the required instrumentation. The synchrotron SAXS/WAXD instrument has three sections: a collimation system, a sample chamber, and a detection section. They need to be linked together. At the same time the built-in flexibility of each section should be independent of each other. Thus, the use and coupling of optical tables can be a reasonable choice, though not necessarily the only choice.

There is a large variety of associated instrumentation that can be coupled to light or/and X-ray probing radiation. Space and stability requirements should be considered in advance. Finally, it is reasonable to try to take advantage of dual usage for instrumentation at the synchrotron facility and for home laboratory use. Thus, the ease on assembly/disassembly and transport of components should also be considered in advance.

Finally, BC has been very fortunate to have many talented students and post-doctorates in his research team. Several have written books and extensive reviews. On the topic of laser light scattering and particle analysis, BC takes this opportunity to mention Walther Tscharnuter, Bruce Weiner, and Ren-liang Xu.

(1) An introductory book (available at Amazon – search under Bruce Bradley Weiner) by Bruce Bradley Weiner, on "Let There Be Light: Characterizing Physical Properties of Colloids, Nanoparticles, Polymers and Proteins Using Light Scattering, ISBN-13: 9780578428574 (2019)" should be of interest. Walther Tscharnuter, together with Weiner, has successfully combined DLS, SLS, electrophoretic light scattering (ELS) and phase analysis light scattering (PALS) in one instrument.

(2) A (2022) book on Optical Technology and Applications of Particle Characterization 颗粒表征的光学技术及应用, by (许人良) Renliang Xu, published by the Chemical Industry Press 化学工业出版社, 412 pp., with *1017 references* (including some references in Chinese). It covers major technologies of particle characterization.

(3) The reader may find a similar set of collection on Quasi-elastic Light Scattering by Macromolecular, Supra-molecular, and Fluid Systems, edited by BC, a SPIE Milestone Series, Volume MS 12, *720 pp*, 1990, where many key publications by established scientists/engineers on the *development* of laser light scattering have been assembled in one volume.

References

Author's Selected Light Scattering Book/Review Publications

1. B. Chu, Laser Light Scattering, *Ann. Rev. Phys. Chem.*, **21**, 145 (1970).
2. B. Chu, *Laser Light Scattering*, Academic Press, New York and London, 337 p (1974).
3. NATO ASI Series B: Physics, Volume 73. *Proceedings on Scattering Techniques Applied to Supra-molecular and Non-equilibrium Systems* (S. H. Chen, B. Chu, and R. Nossal, eds.), Plenum Press, 928 p. (1981).
4. B. Chu, Correlation Function Profile Analysis in Laser Light Scattering. I. General Review on Methods of Data Analysis. In: J. C. Earnshaw and M. W. Steer, eds., *The Application of Laser Light Scattering to the Study of Biological Motion*, Plenum Press, New York, pp. 53–76 (1983).
5. B. Chu, Photon Correlation Spectroscopy of Polymer Solutions. In: C. Price and C. Booth, eds. *Comprehensive Polymer Science: Polymer Characterization*, Vol. 1, Pergamon Press, Oxford, England, Chapter 8, pp. 161–171 (1989).
6. B. Chu, Laser Light Scattering. In: A. R. Cooper, ed., *Determination of Molecular Weight*, John Wiley & Sons, NY, Vol. 103, Chemical Analysis Series, pp. 53–86 (1989).
7. SPIE Milestone Series. Vol. MS 12, *Selected Papers on Laser Light Scattering by Macromolecular, Supra-molecular and Fluid Systems* (B. Chu, ed.) 736 p. (1990).
8. B. Chu, *Laser Light Scattering: Basic Principles and Practice*, 2nd edition, Academic Press, Boston, (1991) 343 p.; reprinted with new preface by Dover Publications, Inc., Mineola, NY (2007) 368 p.
9. B. Chu, Quasi-Elastic Light Scattering. In: R. W. Richards (ed.) *Scattering Methods in Polymer Science*, Ellis Horwood Publishers, Great Britain, pp. 6–25 (1995).

10. B. Chu, D. Fang and Y. Mao, Instrumentation on Multi-Scaled Scattering of Bio-Macromolecular Solutions, *Int. J. Mol. Sci.* (2015), **16**, pp. 10016–10037.

Pinhole Collimation System (2.6.3)

11. B. Chu and B. S. Hsiao, "Small Angle X-Ray Scattering of Polymers" Chemical Reviews, **101**(6) 1727–1761 (2001).

Chapter 3

Evolution of Electrospinning Technology

3.1 Outline

This chapter is concerned with the technical development of electrospinning/blowing technology. In the overview, its relationship with the current theme of condensed soft matter and in the use of nanofiber technology as well as molecular assembly and engineering on applications to liquid filtration is presented.

The emphasis continues to include a combination of university, industry, and government as the broad basis organization because of the complex disciplines, involving chemistry, engineering, physics, and biology. They are also needed to achieve viable solutions on filtration membranes that can help us to find reasonable solutions as related to water (and air) purifications.

In our presentation of this chapter, it is worthwhile to have an overview on the topic of water (and air) purification. We should note that our primary interest in the presentation of this chapter is on the use of electro-spinning/blowing technology as one of the means of producing nanofibers that can be used as a crucial component for the development of water (or air) filtration membranes. More appropriately, the topic of water (or air) filtration, as one of the primary needs for the well-being of our society, has, in fact, a broader perspective in its relationship to the production of *purified water* over a range of diversified

Laser Light and X-Ray Scattering: With Evolution of Electrospinning Technology in Biomedical Applications and Water Purification
Benjamin Chu

ISBN 978-981-4968-88-1 (Hardcover), 978-1-003-54074-8 (eBook)
www.jennystanford.com

applications. Thus, the scientific contributions in this chapter represent only one aspect of the overall theme of providing purified water for the benefits of society. Again, there will also be involvement by the local and national government, as we have previously mentioned in Chapter 2.

The process on the purification of water depends on the nature of impurities and the quality of purified water for different applications. For example, the acceptable impurities for *drinking* water are different from those of *agricultural* water. The overall objective is always closely related to the economic requirements, i.e., the final overall cost for the amount of water produced is based on a specific purpose. Thus, it is often cheaper to purify *waste* water than to purify *sea* water from the ocean. Extensive development on the transport and storage of wastewater, together with wastewater purification, can often be more economical when compared with sea water desalination processes. The city of San Diego (where the author now lives) in California, for example, has undergone some such comparisons. It is interesting to learn how the reservoir construction for the storage of wastewater may include even major highway rerouting, thereby permitting new water ducts for wastewater transport underneath or by pass a multilane highway for access to a large wastewater reservoir connected to a wastewater purification plant. It should be particularly important to understand that the overall cost has to be justified in comparison with a water purification plant by using the desalination of seawater, where clean seawater is available from the nearby coast in California.

Electro-spinning/-blowing technology is not the only pathway to produce nanofibers which are suitable as an effective component in formulating filtration membranes. There are other chemical or/and physical approaches that can produce a low-cost filtration layer for water purification. For example, bioprocessing of plant-based materials, such as cellulose, can produce effective filtration membranes and chemical modifications can be used to create active sites to produce membranes that are able to absorb or reject specific dissolved or suspended components in water in order to yield desirable purified water, whose specifications can meet the intended objective.

It should be noted that the processes under discussion often deal with large-scale operations. The scale-up process should be considered early in the game in order to see whether the new designs will be able to achieve the intended goals. There are numerous reviews and extensive publications or/and patents on this topic. The recommended references will be stated at the end of the chapter and new reviews, e.g., *Accounts of Chemical Research*, shall be forthcoming.

In this chapter, only patents, patent applications and scientific journal publications in early 2000 of *one research team* at Stony Brook University are presented. The purpose of this Inclusion has two reasons: (1) to show there are *extensive* literature and patent information involved, and (2) information related to industrial applications is complex, especially with respect to patents, because there is the technical knowhow which is often hidden and not easily discovered.

The electro-spinning/spraying technology can produce a variety of formats and involves a complex process, with many (coupled) variables. Furthermore, some of the variables are related to hardware designs, in addition to the materials used. Thus, for optimal production, the technology requires a good understanding on the system of interest, the chemical parameters involved, and the trade off on cost versus desired properties for the final, often specific, product.

For electro-spinning/-blowing technology, we start with general considerations on possible solutions to the technical barriers in the process. It is noted that, to our knowledge, the current production rate for nanofibers by means of electro-spinning cannot be further increased substantially. Nevertheless, there are electro-blowing and other types of improvements which may offer reasonable alternatives for faster production rate or/and lower overall cost.

In this chapter, the emphasis is to pay attention on the multiple variables associated with electro-spinning/-blowing processes, including materials and processing variables which are often coupled together in a complex manner. Examples in Chapter 4 are concerned with topics that can broaden the use of nanofibers for applications to other fields, such as biomedical engineering applications where the emphasis is often on smaller sample size with higher precision controls on materials specifications as well as less emphasis on the production cost.

3.2 Condensed Soft Matter

In dealing with nanofiber technology & molecular assembly/engineering for environment & health, the outline of this section can be presented as follows.

3.2.1 Philosophy and Approach
3.2.2 Technical Development on Electro-Spinning Technology
3.2.3 Advanced Characterization Facilities
3.2.4 Applications
3.2.5 Anti-Adhesion Films - Tissue engineering
3.2.6 Functionalization for bone cell adhesion & orientation for heart tissue regeneration
3.2.7 Bone Mineralization Fundamentals, and
3.2.8 Emphasis on Liquid Filtration

The focus of this chapter is electro-spinning/-blowing technology. The examples in Chapter 4 are concerned with topics that can expand the utility of nanofibers for applications to other fields, such as further development toward devices in biomedical engineering.

3.2.1 Philosophy and Approach

The main theme for this chapter deals with the evolution of electro-spinning technology, with applications to biomedical applications and liquid purification processes. Our emphasis is on unique approaches, i.e., this chapter is concerned primarily with electro-spinning technology (Sections 3.2.2 to 3.2.4). The specific targets are related to the use of nanofibers for water purification processes. It should again be emphasized that, in applications associated with water purification, cost-effectiveness in the final application, often dealing with large-scale operations, should be taken into account *early* in the overall process. Furthermore, *plant* sources, instead of petroleum chemical-based materials, for fibers are important considerations.

The approach is generally interdisciplinary in nature, as the topic involves

a. Synthesis	(Chemistry)
b. Processing	(Engineering)
c. Characterization	(Physics)
d. Functions	(Biology & Medicine)

Emphasis on using new materials (chemistry), with processing by means of electro-spinning technology (engineering), and characterizations (physics), for applications of benefits to society (e.g., biology, water, and air) represents the key disciplines of this chapter.

3.2.2 Technical Development on E-Spinning Technology

The **Chemistry** involved has a broad scope and we shall defer the discussions to specific cases. We start with **Processing (Engineering).** However, the processes also cover a broad range of technology, with materials for use in

a. Membranes:	Electro-spinning/electro-blowing technology and gel/melt spinning for the production of small diameter fibers as well as layer by layer processing
b. Films:	Spin casting
c. Particles:	Hydro-gels; nanometer (nm)-size to μm-size particles
d. Foams:	Micro-phase separation in liquid/solid phase transitions by super-critical fluid processing, by crystallization, and by salt substitution

Our emphasis will be on electro-spinning/electro-blowing technology and for fiber processing by using gel/melt spinning.

In Section 3.2.1.C "Characterization (Physics)," a broad range of techniques is involved. The overall instrumentation can be subdivided into 4 different types of physical measurements, as summarized below:

1. Microscopy, including, for example, electron microscopy, atomic force microscopy, and optical microscopy;
2. Scattering, including laser light scattering, SAXS/WAXD where synchrotron facilities now become a topic of interest, and small angle neutron scattering (SANS);
3. Spectroscopy, including Raman scattering, FTIR, and NMR;
4. Thermal and mechanical measurements, such as calorimetric, rheological, and stress-strain measurements.

Aside from scattering, the other techniques have been included to show the broad background and the reason for our restrictions to the current topics in the discussion.

For background information, patents and patent applications for multiple-jet electro-spinning and biomedical applications over a limited time period are presented. It is intended to show to the reader the importance of team experience and the scope of involvement in the topics under the current discussion for this chapter.

Selected patents and patent applications for multiple-jet electro-spinning and biomedical applications from the Chu labs are listed as follows:

- B. Chu, B. S. Hsiao, D. Fang and C. Brathwaite, "Biodegradable and/or Bio-absorbable Fibrous Articles and Methods for Using the Articles for Medical Applications." U.S. Pat. 6685956, 2004.
- B. Chu, B. S. Hsiao, D. Fang and C. Brathwaite, "Biodegradable and/or Bio-absorbable Fibrous Articles and Methods for Using the Articles for Medical Applications." U.S. Pat. 6689374, 2004.
- B. Chu, B. S. Hsiao and D. Fang, "Apparatus and Methods for Electro-spinning Polymeric Fibers and Membranes." U.S. Pat. 6713011, 2004; PCT Int. Appl. WO 0292888.
- B. Chu, B. S. Hsiao, M. Hadjiargyrou, D. Fang, S. Zong and K. S. Kim "Cell Delivery System Comprising a Fibrous Matrix and Cells," U.S. Pat. 6790455, 2004.
- B. Chu, B. S. Hsiao, D. Fang, Akio Okamato "Electro-Blowing Technology for Fabrication of Fibrous Articles and Its

Applications of Hyaluronan," U.S. Pat. Appl. Serial No. 10/674464, 2003.
- B. Chu, B. S. Hsiao, D. Fang and K. S. Kim, "High Flux and Low Fouling Nanofibrous Ultrafiltration Membranes," Filed by SUNY-Stony Brook (R-7760), October 6, 2004.
- B. Chu, B. S. Hsiao, D. Fang, Akio Okamato, "Cross-linking of Hyaluronan Solutions and Nanofibrous Membranes Made Therefrom," Filed, August 27, 2004.
- B. Chu, B. S. Hsiao and D. Fang, "Apparatus for Electro-Blowing or Blowing-Assisted Electro-Spinning Technology and Process for Post Treatment of Electro-spun or Electro- blown Membranes," Filed, September 9, 2004.

Selected Publications of Electro-Spun Nanofiber Scaffolds in Scientific Journals (2002 to 2004)

1. Xinhua Zong, Dufei Fang, Kwang-Sok Kim, Jeyoung Kim, Sharon Cruz, Benjamin S. Hsiao and Benjamin Chu, "Structure and Process Relationships in Bioabsorbable Nano-fiber Membranes by Electro-spinning," *Polymer*, **43**(16), 4403–4412 (2002).
2. Xinhua Zong, Kwangsok Kim, Shaofeng Ran, Dufei Fang, Benjamin S. Hsiao and Benjamin Chu, "Structure and Morphology Changes during In Vitro Degradation in Electro-spun Poly(glycolide-co-lactide) Bio-absorbable Nano-fiber Membranes," *Biomacromolecules*, **4**(2), 416–423 (2003).
3. Y. K. Luu, K. Kim, B. S. Hsiao, B. Chu, M. Hadjiargyrou, "Development of a Nano-structured DNA Delivery Scaffold via Electro-spinning of PLGA and Block Copolymers," *J. Control Release*, **89**, 341–353 (2003).
4. Xinhua Zong, Shaofeng Ran, Benjamin S. Hsiao and Benjamin Chu, "Control of Structure/Morphology and Property of Poly(glycolide-co-lactide) Nano-fiber Membranes via Electro-spinning and Post-treatments," *Polymer*, **44**(17), 4959–4967 (2003).
5. Kwangsok Kim, Meiki Yu, Steven X. Zong, Jonathan Chiu, Dufei Fang, Young Soo Seo, Benjamin S. Hsiao, Benjamin Chu and Michael Hadjiargyrou, "Control of Degradation Rate and Hydro-philicity in Electro-spun Poly(DL-lactide) Membranes for Cell Scaffolding," *Biomaterials*, **24**, 4977–4985 (2003).

6. Xinhua Zong, Sean Li, Elliot Chen, Barbara Garlick, Kwangsok Kim Dufei Fang, Jonathan Chiu, Thomas Zimmerman, Collin Brathwaite, Benjamin S Hsiao, and Benjamin Chu, "Prevention of Post- Surgery Induced Abdominal Adhesions by Novel Electro-spun Bio-absorbable Nano-fiber Membranes of Poly(lactide-co-glycolide)," *Annals of Surgery*, **240**(5), 910–915 (2004).
7. Hongliang Jiang, Dufei Fang, Benjamin S. Hsiao, Benjamin Chu and Weiliam Chen, "Optimization and Characterization of Dextran Membranes Prepared by Electro-spinning," *Biomacromolecules*, **5**(2):326–333 (2004).
8. In Chul Um, Dufei Fang, Benjamin S. Hsiao, Akio Okamoto and Benjamin Chu, "Electro-Spinning and Electro-Blowing of Hyaluronic Acid," *Biomacromolecules*, **5**(4), 1428–1436 (2004).
9. Kwangsok Kim, Yen K. Luu, Charles Chang, Dufei Fang, Benjamin S. Hsiao, Benjamin Chu and Michael Hadjiargyrou, "Incorporation and Controlled Release of Hydrophilic Antibiotics Using Poly(lactide-co-glycolide) Based Electro-spun Nano-fibrous Membranes," *J. Controlled Release*, **98**(1) 47–56 (2004).

It should be noted that presentations to conferences and preprints, e.g., Xinhua Zong, Dufei Fang, Kwangsok Kim, Shaofeng Ran, Benjamin S. Hsiao, Benjamin Chu, Collin Brathwaite, Sean Li and Elliot Chen "Nonwoven Nanofiber Membranes of Poly(lactide) and Poly(glycolide-co-lactide) via Electro-spinning and Their Applications for Anti-Adhesions," *ACS Polymer Preprints*, 2002; **43**(2):659–660, have been omitted. It should also be of interest to the reader that there were many other activities during the time period of 2002–2004. In particular, the publications of Seeram Ramakrishna at the National University of Singapore should be of interest. One highly cited review on polymer nanofibers by electro-spinning and their applications in nanocomposites (ZM Huang, YZ Zhang, M Kotaki, S Ramakrishna) in *Composites Science and Technology* **63**(15), 2223–2253 (2003) should be worthy for the reader to take a close reading.

With the above background information, the main tasks of this chapter are presented below. The primary intention is to provide the reader, who is not physics/engineering oriented, a physical picture for the basis of electro-spinning technology.

Thus, the discussions have emphasized on the *physics* associated with the controlling variables as related to the electro-spinning process, leaving the detailed *mathematical equations* and future development to those who are interested in advancing the technology further.

The formation of nanofibers from a polymer solution starts with the polymer solution leaving the *tip of the spinneret*. The polymer fluid droplet under the influence of gravity and of electric field, as shown schematically in Figure 3.1, reveals a range of effects that we need to consider. From gravity and if the spinneret is pointing downward, the extrusion forces depend on the surface tension of the polymer solution, while the surface tension depends on the polymer solution concentration and additives which we may add to the polymer solution so as to partially control the electro-spinning process. At the same time, we can affect the electric field strength: 1. by changing the shape of the electrode, 2. by changing the electric field distribution, and 3. by adding excess charges to the polymer solution. When the polymer solution leaves the tip of the spinneret, it is being stretched.

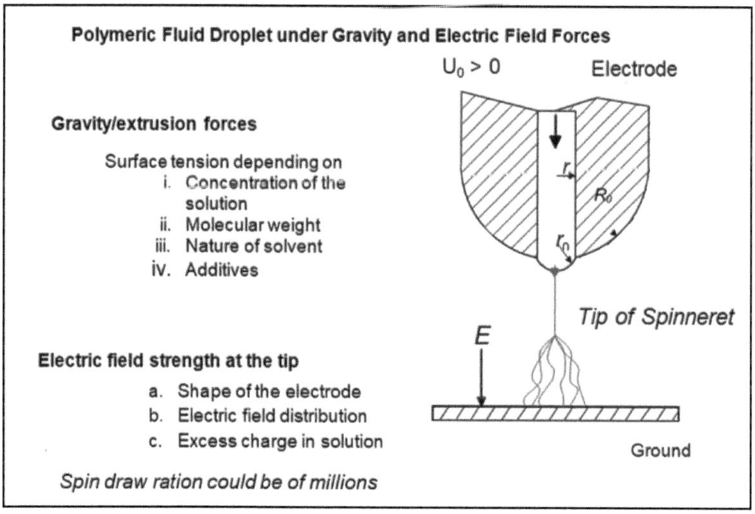

Figure 3.1 Fundamentals on electro-spinning technology.

A more detailed picture is being shown in Figure 3.2 on the fiber formation by the polymer solution stream after it has left the spinneret. The photograph on the left shows the polymer solution stream moving in an electric field E, being pulled and developed instability. A cartoon to the right depicts how the polymer chains are being pulled. In the meantime, the polymer solution stream undergoes solvent evaporation, the electric charges undergo repulsion, and the stretched polymer chains want to contract back like a spring. Soon the polymer solution jet develops instabilities, as shown by the photograph on the left.

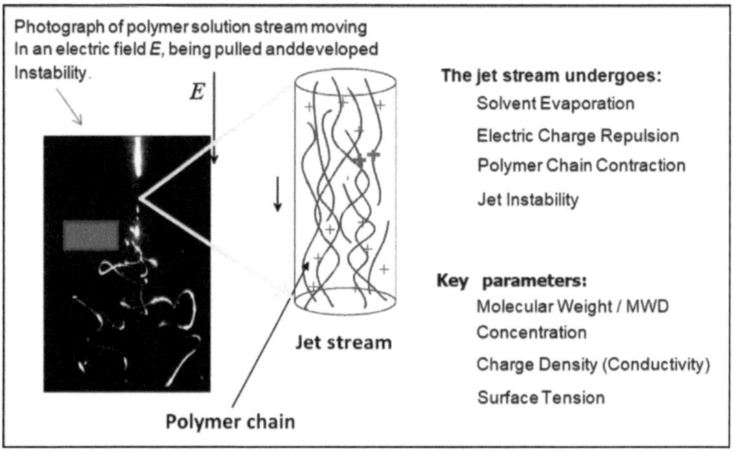

Figure 3.2 Electro-sspinning process: fiber formation, after the jet stream has left the electrode.

While the key parameters can be stated as polymer molecular weight and its molecular weight distribution, polymer concentration in the solution, the charge density in the jet stream, and the surface tension of the polymer solution, all those parameters are changing with time and the changes are interrelated. Furthermore, as the solvent is being evaporated in this process, all those interrelated parameters are changing. Thus, we can see that electro-spinning is a complex time-dependent process.

Our intention, as has been mentioned briefly before in Chapter 1, is to emphasize the physics, leaving the mathematical equations aside. Later on, we shall introduce one figure with some mathematics to form a bridge for readers who may want to learn electro-spinning at a deeper level. The qualitative approach below is intended for the reader to see what the variables are in the electro-spinning process so that when we use such an instrument, we can take advantage of the instrument to achieve the intended results.

While we are dealing with the formation of nanofibers as a crucial component for filtration membranes used in water purification processes, there are other applications. For example, as we have mentioned it before, air filtration immediately comes to our mind. The demonstration comes to mind, is shown in the next figure, which appeared in *National Geographic*, January 2003, pp. 52–53, as a demonstration by the U.S. Army Natick Soldier Systems Center in Natick, MA.

It is interesting to know that the use of nanofibers for more effective face masks started more than 20 years ago. Historically, electro-spinning was first observed by Rayleigh in 1897 and studied in detail by Zeleny in 1914. Electro-spinning is a complex process that has been around for some time. Due to its complexity, we shall try to emphasize the variables associated with the processes in a qualitative manner, leaving the quantitative mathematical details aside. Nevertheless, some details on the coupled variables during processing are presented to show the complexity of the many variables associated with such a process.

National Geographic, January 2003, pp. 52–53

In the January 2003 issue of *National Geographic* (pp. 52–53), we have noted that the Army Natick Laboratory had already started experiments on the use of nanofibers for more effective face masks. A closer look at the structures in electro-spun/sprayed biodegradable polymers (PLA, PGA), using different concentrations of polymers and operating conditions can produce polymer micro-structures from nanoparticles to nanofibers. Figure 3.3 demonstrates the flexibility of the electro-spinning/blowing technology.

There are many different ways to alter the electric field distribution. We shall provide a few additional examples to illustrate the diversity of this approach. Thus, complexity comes with flexibility which is useful in trying to create new structures for applications to other topics, such as biomedical applications, a topic with considerations where useful functions may supersede financial cost as the most important consideration. Further examples will be provided in Chapter 4 on applications.

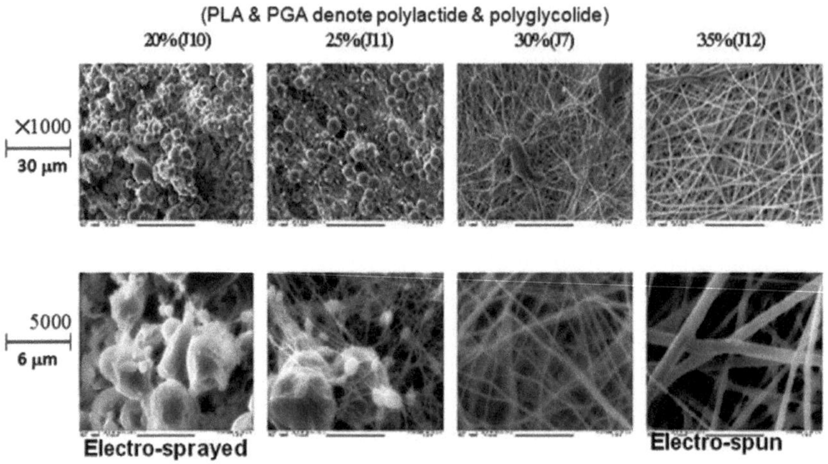

Figure 3.3 Attainable structures in electro-spun/sprayed bio-degradable polymers (PLA, PGA) from nanoparticles to nanofibers.

We shall pay attention to the effect of concentration which alters the solution viscosity during the electro-spinning/blowing process. For further details, please see the reference: Xinhua Zong, Dufei Fang, Kwang-Sok Kim, Jeyoung Kim, Sharon Cruz, Benjamin S. Hsiao and Benjamin Chu, *Polymer*, 2002; **43**(16): 4403–4412.

With the information presented above, it becomes feasible to consider a non-woven nanofiber platform where electro-spinning/-blowing technology can play a very useful role in the production of nanofibers for a range of applications. The current objectives have been limited to:

- Biomedical Applications
- Filtrations (Air and Fluid)
- Technology Development – to simplify and to expand the process in combination with other preparation methods.

We recognize the existence of a core technology and fundamental science that support the above objectives:

- Nanofiber Processing (*Electro-spinning* and *Electro-blowing*)
- Polymer Synthesis
- Structure/Property Optimization
- Functionality (e.g. Controlled Delivery of Drugs)

Thus, knowledge in chemistry, physics, engineering, and biology remains the basic foundation of our endeavor. Discussions presented in Chapter 4 continue to be relevant.

3.2.2.1 Uniqueness of es-Jets™ technology

The aims of es-Jets™ Technology (Figure 3.4) are as follows:

- Control of fiber diameters (down to ~10 nm dia.)
- Control of membrane porosity
- Control of chemical composition of fibers
- Pattern formation
- Multi-layer formation
- Multi-jet technology for scale up operation

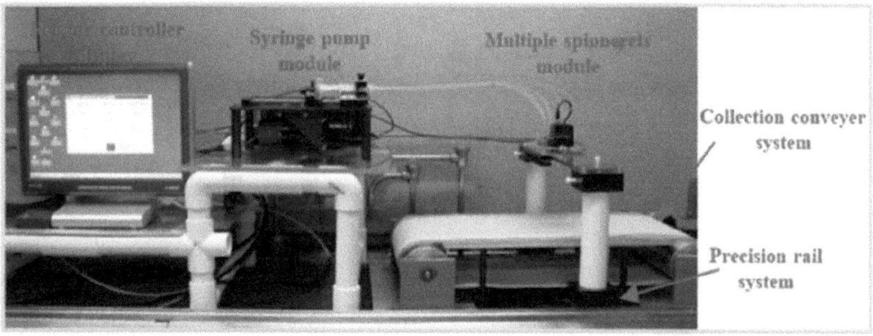

Figure 3.4 es-Jets™ technology: B. Chu, B. S. Hsiao and D. Fang, U.S. Pat. 6713011 (2004); PCT Int. Appl. WO 0292888.

The small-scale experimental setup of the es-jets™ technology has been tested for the capabilities as presented by using a range of polymers. The key components are related to the use of multiple spinnerets and a collection conveyer belt system with a precision rail to provide control of the electro-spinning process.

For the user of such an instrument, considerations on the electro-spinning technology as related to the gravity/extrusion forces and electric field strength at the tip as well as the key parameters, such as molecular weight/MWD, concentration, charge density of the jet, polymer concentration and surface tension of the fluid at the tip of the spinneret have been presented in two earlier figures. The primary objective for this chapter is to provide the reader with an overview on the electro-spinning process, based primarily on the *physics* associated with the process, instead of more quantitative *mathematical equations.* Most readers are interested in using the instrument. Thus, it is of interest to the reader to learn how such experiments should be carried out. The next four figures are intended to provide a reasonable pathway for readers who want to use this type of instruments, primarily being developed by others, such as a commercial instrument. We need to pay attention on the variables associated with the electro-spinning process.

In Figure 3.5, material (yellow) and processing (light green) variables have been separated. Yet we know that those variables are *inter-connected.*

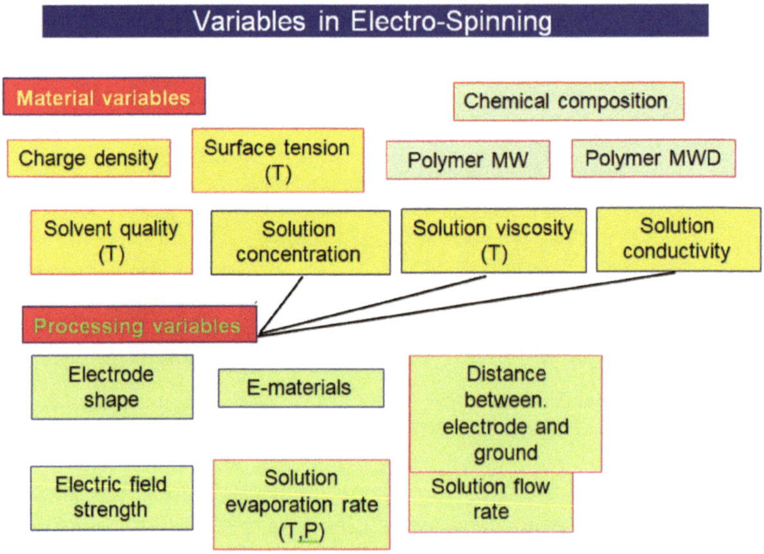

Figure 3.5 Variables in electro-spinning.

Take note of the variables involved, starting with Figure 3.5. It should be noted that the variables are interconnected making electro-spinning a complex process. Here, T, P, MW, and MWD denote temperature, pressure, molecular weight, and molecular weight distribution, respectively.

We should also note that the initial jet formation starts when the extrusion force is greater than the retention force, meaning that the net force (Fs) = $\rho_s(\sigma) E > [Y - \rho_0 Vg]$, as shown in the next figure on variables in electro-spinning (Figure 3.6).

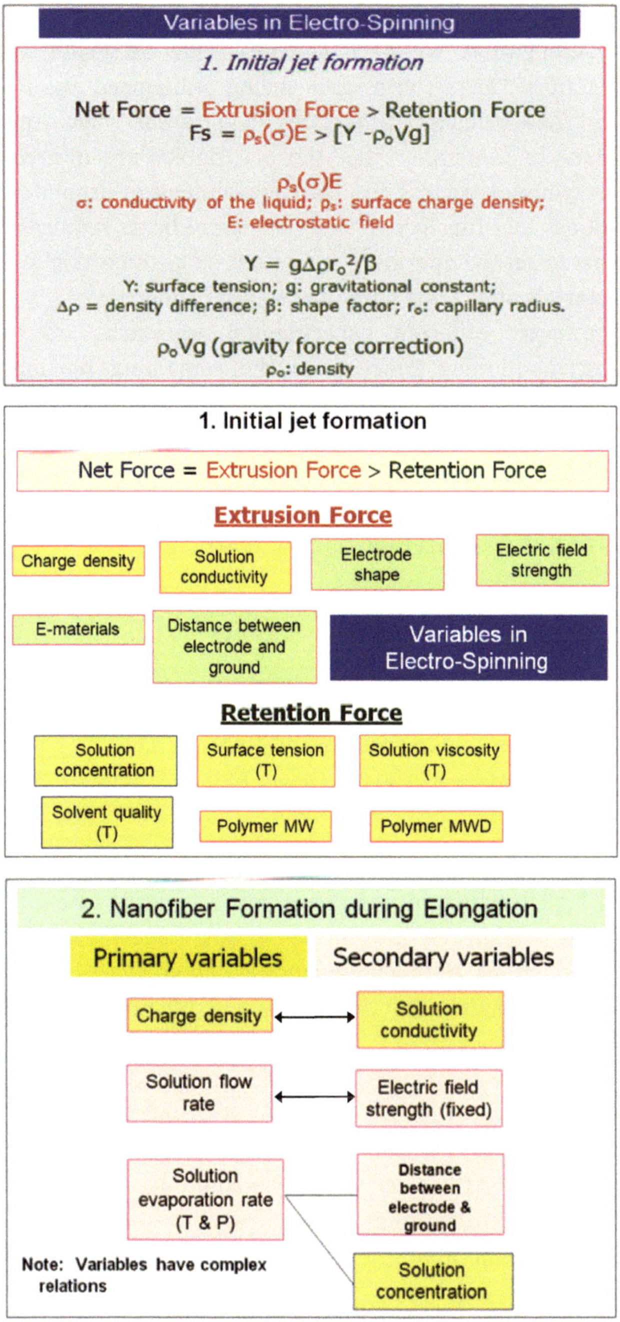

Figure 3.6 Variables in electro-spinning.

In the (1) initial jet formation and (2) nanofiber formation during elongation, we have separated the extrusion force and the retention force with each being influenced by materials, physical, and electrical properties in different ways. Again, it is important to remember that those variables are interconnected in their complex ways. Thus, the development of proper operating conditions for the processing of nanofibers requires careful thoughts to set the operating conditions so as to be able to produce the materials of interest with the desired properties.

There are different experimental approaches to influence the electro-spinning process. In the remaining portion of the chapter, the presentation will provide illustrations on a few examples of modifying different variables so as to alter the electro-spinning process, making the technology to enable us to reach the final goals of producing nanofibers with desirable properties. The example modifications are as follows.

- Introduction of secondary electrode so as to be able to influence the solution jet behavior
- Use of multiple jets to increase the production rate
- Adding gas (mostly air) streams to modify the jet stream, known as blowing assisted electro-spinning
- Use of electric field-assisted blowing, known as electro-blowing.

We know that the electro-spinning/blowing processes do not have a high production rate. Thus, for large scale operations, it is important to consider alternative routes. Nevertheless, for small scale productions, the technology offers good control on nanofiber formation, implying that its utility for biomedical applications should not be ignored.

One method to change the electric potential distribution of an electrode at the tip of the spinneret is to introduce a secondary electrode. The figure with multiple jets in the presence of secondary electrodes shows the electric field distributions as being computed by using a large scale finite element analysis.

The component of the secondary electrodes for a multiple spinneret assembly is shown in the next figure. It should be noted that there are different pathways to change many of the variables in the materials used, the processing variables, and the environment of the setup, just to name a few. Thus, depending on the viewpoint, knowledge, experience, and cost considerations, the complexity also offers opportunities.

3.2.2.2 Electric potential distribution of multiple Jets

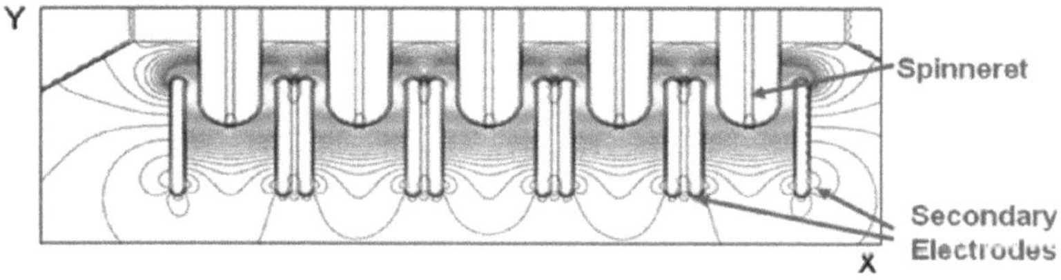

Figure 3.7 Large-scale finite-element analysis calculation.

Optimize:

- Electric field strength on the tip
- Separation distance of the electrodes
- Relative position of electrode elements
- Relative electric potentials on the electrodes

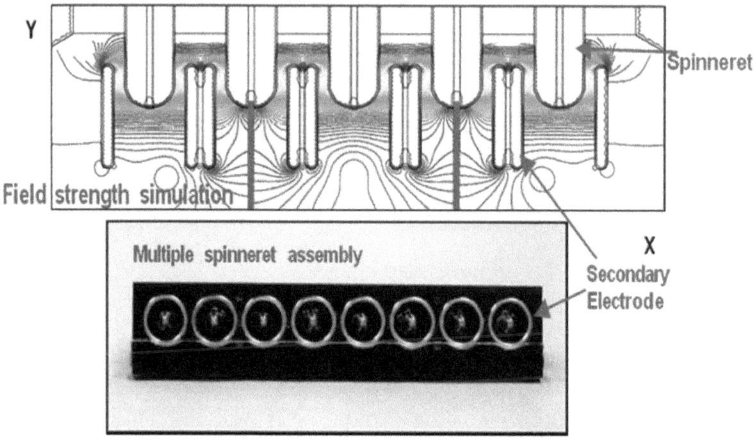

Figure 3.8 Component: secondary electrodes.

We have presented one approach on the electric field modification by using secondary electrodes with a multiple spinneret assembly. There are undoubtedly different ways to modify the applied electric field profile and thereby the exit jet stream at the tip of the spinneret. The worth-while consideration should always be associated with the overall process, including spinneret cleaning.

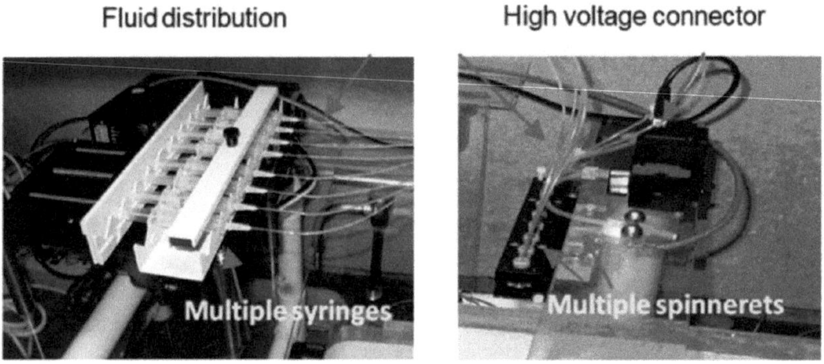

Figure 3.9 Components of multiple-jet device. Note: shared components in eletro-spinning with/without gas-flow assistance.

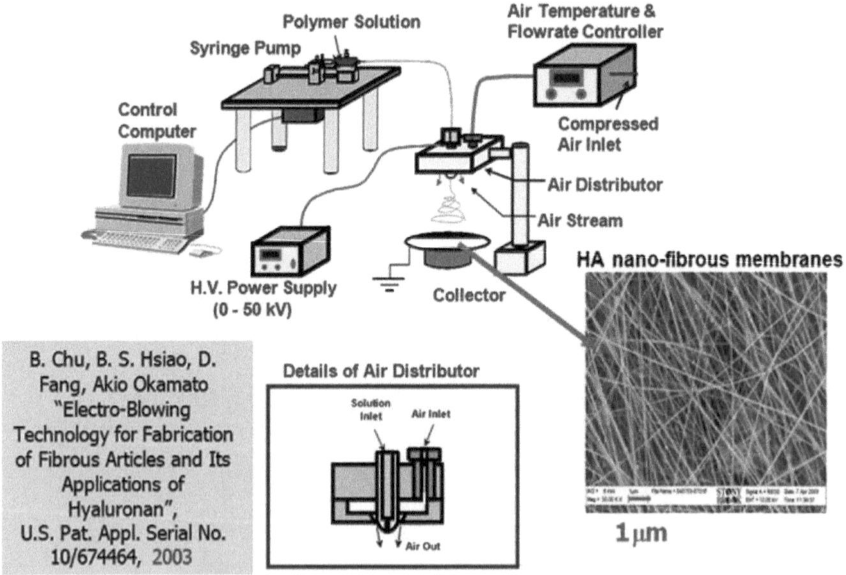

Figure 3.10 Blowing-assisted electro-spinning.

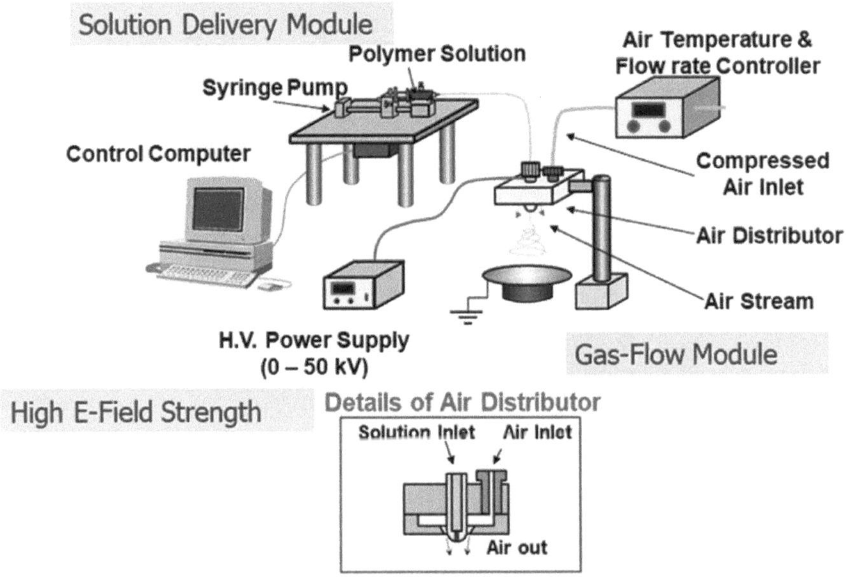

Figure 3.11 Electro-blowing/spinning device.

3.2.3 Blowing-Assisted Electro-Spinning vs. Electro-Blowing (Electric Field-Assisted Blowing)

- Both processes consist of the use of electric force and gas-blowing shear force.
- In the blowing-assisted electro-spinning process, the electric force is the dominating factor. The gas-blowing feature can assist in shearing the fluid jet stream and in *controlling* evaporation of the solvent. (Lower throughput, smaller diameter)
- In the electro-blowing process, the gas-blowing force is the dominating factor in the initial stage to achieve the desired spin-draw ratio, while the electric force may enable the further elongation of fiber (higher throughput, larger diameter).

Electro-blowing technology (in general) is referred to as EBT.

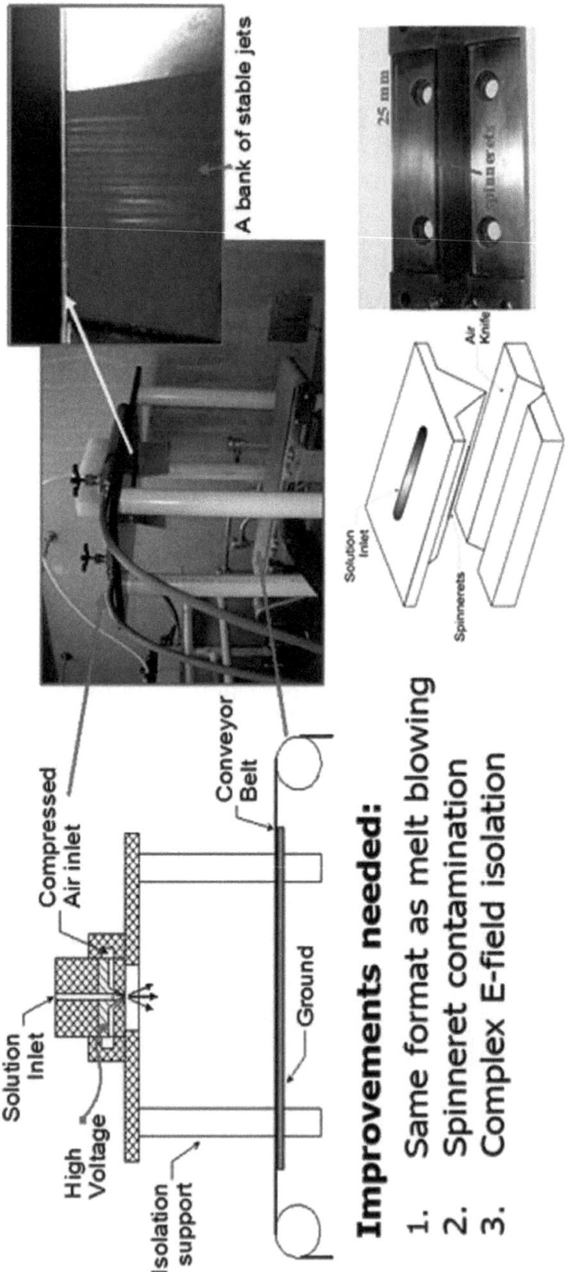

Figure 3.12 Schematic and photographs of multiple-jets electro-blowing device on an isolated platform.

Figure 3.12 explains how such an apparatus works. It is particularly important to pay attention to the solution inlet and the spinneret location, as shown by a combination of the schematics and the picture on the lower right corner of this presentation.

The improvements (1–3) to which we refer in Figure 3.12 imply that such a setup can be used (1) for melt blowing, (2) feature for spinneret cleaning has not been implemented, and (3) complex E-field isolation will be presented below.

An enlarged figure for solution-blowing of a bank of stable jets, as shown in the upper right corner of Figure 3.12, is presented in Figure 3.13. More details on electro-blowing shall be presented below.

Figure 3.13 Multiple-jets solution-blowing without electric field.

3.2.4 Major Technical Barriers and Solutions for Manufacturing Nanofibers by Electro-Spinning

Current Technical Barriers

- Low speed of fabrication
- Limitation of process to polymer solutions
- Lack of ability for sustained long-term operation

Key Points: We need (1) Robust Operation, (2) Quality Control, and (3) Cost-effective Solutions. To overcome those technical barriers, we consider the following pathways.

- Combination of applied electric field & flowing gas stream
- Use of multiple jets
- Simplified electric-field isolation & controlled electric-field distribution
- Ease of cleaning or Self-cleaning of spinnerets
- Controlled evaporation/solidification processes

The solutions have been discussed in our 2004 patents and patent applications.

Note: The presentation above illustrates the multi-parameters involved in the electro-spinning process. Thus, the design requires optimizations which are system and materials dependent.

3.2.5 Example on Production Rate in Using Mechanical Gas Shearing and Elongation as well as Drag Forces

Conditions

Spin-draw ratio ~ one million

Example, 10 (w/v)% polymer solution producing 0.8 g/20 hour/spinneret; Spinneret cross-sectional area = 0.04 mm^2.

Estimated Parameter Values

Initial fluid velocity = ~10 m/hour.

With spin-draw ratio = 10^6, final fiber cross-sectional area = 0.04 μm^2

Or a fiber diameter of ~100 nm (slight over-estimate);

Remembering 90% of polymer solution is the solvent, the final fiber speed reaching the collector could be estimated to be ~1,000 km/hour, or ~600 miles/hour, approaching the sound velocity.

Thus, the production rate per spinneret cannot be increased much higher.

3.2.6 Parameter Considerations for Further Improvements

Self-Cleaning Options

- Use of large spinneret apertures
- Fluid flow controlled by tapered pin to vary the gap size

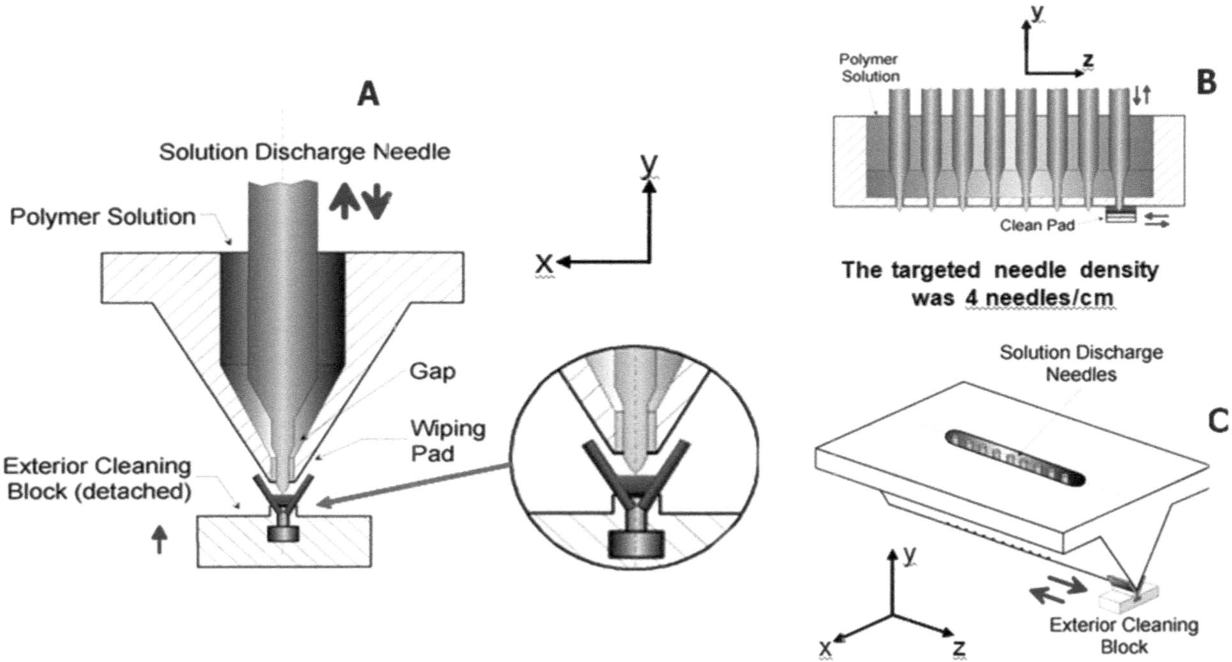

Figure 3.14 Designs of internal cleaning (A), with needle assembly (B), and of external cleaning (C) for multiple-jets electro-blowing operation.

- Pin position controls fluid channel size
- Facilitate cleaning of fluid channel, tapered pin, outside of spinneret aperture hole
- Tapered pin acting as an electrode

Control of Solvent Evaporation

- Temperature control along jet stream flight path
- Gas flow versus solvent removal

Electric Field Reversal

The above considerations have provided us first-order pathways for further improvements of the electro-spinning/blowing processes. Some details on the implementations are presented as follows.

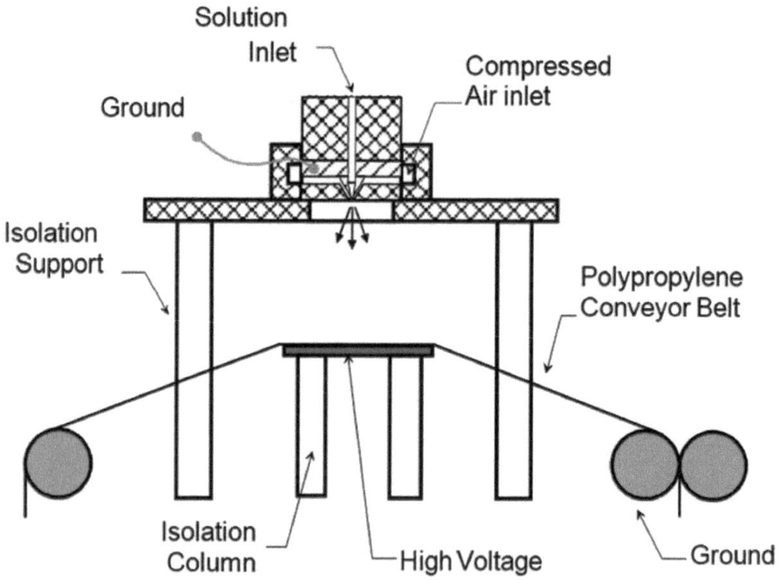

Figure 3.15 Reversal of electric field polarity.

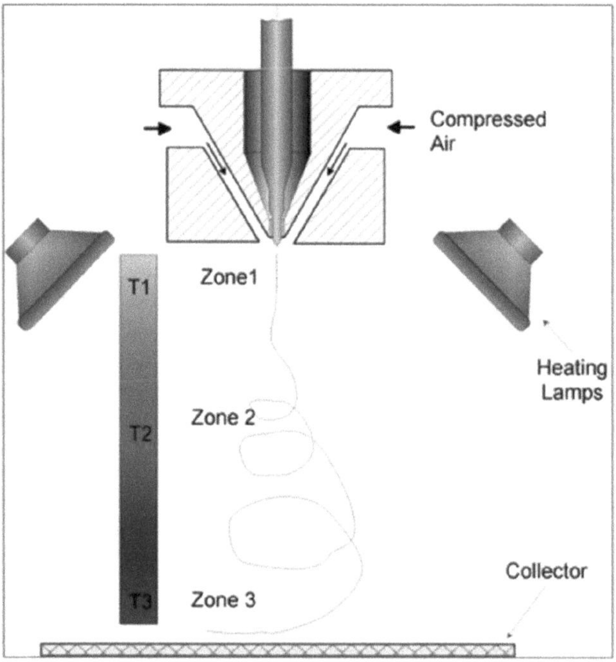

Figure 3.16 Schematic diagram for control of evaporation/solidification.

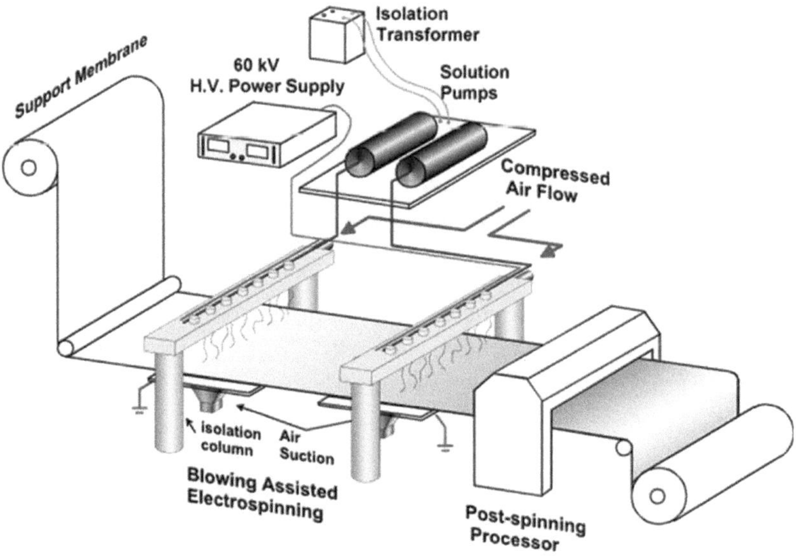

Figure 3.17 Multiple-jets blowing-assisted electro-spinning.

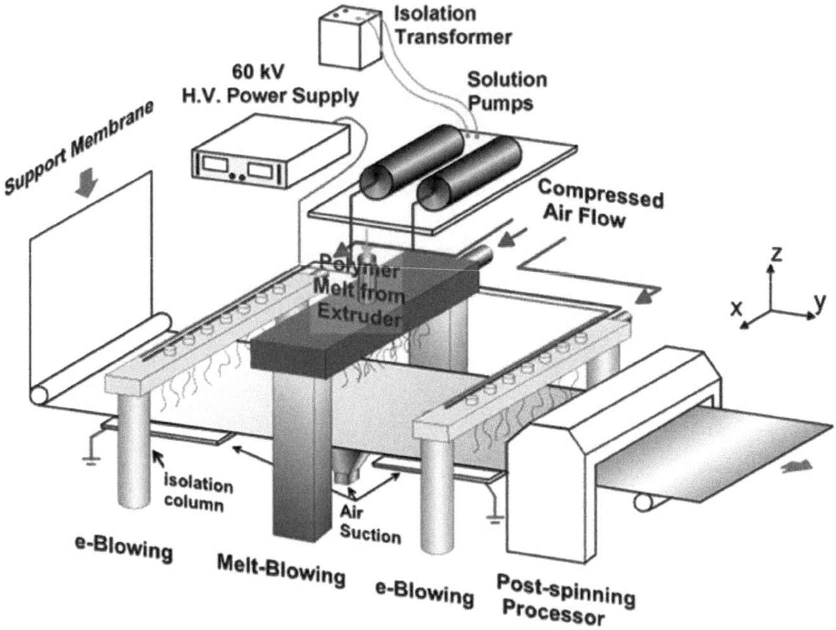

Figure 3.18 Schematic diagram of hybrid process of multiple-jet blowing-assisted electro-spinning or electro-blowing with melt-blowing.

3.3 Conclusion and Path Forward Discussions

Success depends on exploitation of new ideas to solve problems (and to create products) within the team capability.

Some measurements of innovation and productivity:

1. Successful funding (e.g., NIH SBIR I & II, DoE SBIR)
2. Key individual leadership
3. Talented researchers with multidisciplinary approaches, including collaborative networks
4. First-class infrastructure (facilities) support
5. Strategic planning with risk management

In this chapter, we have presented electro-spinning/blowing technology, in terms of its complexity, limitations, and utility for applications to water (and air) purification. The complex processing approaches suggest the diversity of the technology.

Thus, in biomedical engineering applications, it offers an attractive potential for the delivery of medications that can be tailor designed. This added flexibility is very useful. The variety of specific modifications on the nanofibrous membranes, such as changes in porosity, fiber diameter, surface interactions, chemical/physical adhesion of medications, can offer possible specific products to satisfy the stringent requirements, often demanded for biomedical applications. Again, complexity brings more potential pathways. Thus, it is a two-edged sword depending on how we are able take advantage of this technology.

The reader is being reminded again that while we need to know the fundamentals, it is also important to know where and who the experts are, so that one can work as a team to solve more complex problems by utilizing a range of different disciplines and experiences. At the same time, we should be aware of being able to make specific contributions to solve a complex problem. It is important to know that to take without to give will not form good team work.

In the discussions above, we have not touched upon the approach of using "*Needleless Electro-spinning of Nanofibers*." It is an electro-spinning process of nanofibers without using spinnerets, thereby by passing the difficulties of spinnerets plug up, etc. The approach uses jets induced directly from an open liquid surface. Therefore, it has become a promising technology to produce nanofibers on a large scale. Considerable progress has been made in the development of *needleless electro-spinning*. The reader is being referred to the book *Needleless Electro-spinning of Nano-fibers Technology and Applications*, by Xungai Wang and Tong Lin, ISBN: 9789814316354, published in November of 2013, 204 pp, by Jenny Stanford Publishing Pte. Ltd., Singapore 038988. Web: https://www.jennystanford.com/, for further details. For large-scale operations, such as water (air) purification, it will be a technology that we should not ignore. For example, "Mass production of nano-fibers from needleless electro-spinning by a novel annular spinneret," by Liang Wei et al. in *Materials & Design*, **179**, 107885 (October, 2019), represents one such example.

There is also "A mini-review: Needleless electro-spinning of nano-fibers for pharmaceutical and biomedical applications," by Ioannis Partheniadis, Ioannis Nikolakakis, Ivo Laidmäe, and Jyrki Heinämäki, *Processes*, **8**, 673 (2020); doi: 10.3390/pr8060673.

At present, about half a dozen industrial companies for needleless electro-spinning are working on various aspects of electro-spun fibers for a range of applications. Their contact information suggests locations mostly in the United States, Germany, Italy, and United Kingdom. The instrumentation ranges from laboratory units to large scale industrial operations. Needle-free production becomes especially worthy of attention when dealing with the water (air) filtration process.

A listing of companies with emphasis on needless electro-spinning is presented below.

1. Inovenso (https://www.inovenso.com/) provides unique laboratory/industrial equipment for the production of nanofiber products by using electro-spinning.
2. Elmarco (https://www.elmarco.com) provides needle-free electro-spinning equipment for experimental work in nanofibers, equipped with the same electrode system as used in industrial production lines.

   Nanospider™ technology is a patented, needle-free high voltage, free liquid surface electro-spinning process. The technology is based upon the discovery, that it is possible to create Taylor Cones and the subsequent flow of materials not only from the tip of a capillary, but also from a thin film of a polymer solution.

   - High productivity and scalability
   - High fiber diameter and web uniformity
   - Economical operation and easy maintenance
   - Flexibility in used polymers and substrates

3. Linari Nanotech s.r.l. (info@linarinanotech.com) is a company born as a branch of Linari Engineering, founded in Pisa 2003 by Eng. Stefano Linari.
4. **Fluidnatek®** and **Spinbox®** are the brand names for the two branches of electro-spinning & electro-spraying equipment designed by **Bioinicia,** a company with R&D and production expertise in both electrospinning and

electrospraying processes. Scientists, engineers, project managers and manufacturing operators all working towards a common goal: to support partners at all stages of their development.

5. The Electrospinning Company Ltd (info@electrospinning.co.uk) designs, develops and manufactures nanofibrous biomaterials for use in tissue-regenerative devices.
6. Nanolayer: NanoLayer is certainly one of them. With their proprietary 'sonic' electro-spinning technology, they are capable of producing nanofiber products on a massive scale.

Finally, the reader may be curious and interested to look at images of figures illustrating electro-spinning/blowing (and spin-less nanofiber formation) technology by using Google Search. There are, surprisingly, more than 500 figures (with some forms of overlap) from the literature.

BC leaves three important contributions at the end of this chapter for the reader. It deals with contributions from the leaders in the current topic. They are worthy references for further reading on more details.

1. Seeram Ramakrishna at the National University of Singapore. Review article:

   A review on polymer nanofibers by electro-spinning and their applications in nanocomposites, by ZM Huang, YZ Zhang, M Kotaki, S Ramakrishna, *Composites Science and Technology*, **63** (15), 2223–2253 (2003). BC has modified a very informative figure from this review, which will be used later in Chapter 4.

2. Menachem Elimelech at Yale University in New Haven, Connecticut. The article: Science and technology for water purification in the coming decades, by MA Shannon, PW Bohn, M Elimelech, JG Georgiadis, BJ Marinas, and AM Mayes, *Nano-Science and Technology: A Collection of Reviews from Nature Journals*, published in 2010.

3. Nanocellulose for sustainable water purification, by R Das, T Lindström, P R Sharma, K Chi, and BS Hsiao, *Chemical Reviews*, **122,** 9, 8936–9031 (2022) https://doi.org/10.1021/acs.chemrev.1c00683

It is noted that the book title is *Laser Light and Synchrotron X-Ray Scattering: With Evolution of Electro-Spinning Technology*. Thus, this chapter represents the main focus of the application topic for the book. The broader scope of *plant* fibers implies that, in addition to *ultrafine fibers*, there are fibers that can be obtained from nature, such as cellulose fibers, that can be extracted from plants without the use of electro-spinning/blowing technology. Those *fibers* from nature, with a much broader range of fiber diameters can be utilized for water (air) purification. The discussions start in Section 4.4 (Figure 4.44 to 4.47), where cellulose and chitin nanofibers/whiskers with nominal diameters of ~5–20 nm could be obtained. The review on nanocellulose for sustainable water purification (publication no. 3 above) covers further details.

In dealing with nanocellulose materials, it may be of interest for the reader to be aware of the works involving Alain Dufresne, starting with three books:

1. *Nanocellulose: From Nature to High Performance Tailored Materials* (2012)
2. *Biopolymer Nanocomposites: Processing, Properties, and Applications of Biopolymer Nano-Composites* (2013)
3. *Polysaccharide-Based Nano-Crystals: Chemistry and Applications* (2014)

Thus, researchers started to consider the use of natural plant materials more than a decade ago. Those activities have continued to the present day. For example, Dufresne and coworkers have presented their characterizations of porous date palm fiber in the *Journal of Natural Fiber*, published online in October 2022.

Chapter 4

Potential Applications of Nanofibers

4.1 Brief Summary

Potential applications of nanofibers (see the following text) cover a very broad range. In the current context, liquid filtration for water purification is the second follow-up theme of this book. The applications on how the filtration membranes are being used will be deferred because such examples are system specific, i.e., applications on membrane filtration depend on specific impurities and amount(s) to be removed. Instead, an example of an application for a biological system is presented, partially for bringing to the reader a broader perspective on nanofiber membrane *applications*. The outline on water challenge, as being discussed, points out the unique characteristics of high porosity and inter-connectivity (see Section 4.1.2) of those nanofiber membranes which are excellent as a mid-layer support in many different types of filtration membranes. A range of polymers has been tested as an illustration on the diversity of fibrous membrane systems. Future applications shall certainly involve a broader range of materials with unique filtration properties to cover additional applications to separation technology and biomedical applications.

In this chapter, a qualitative explanation is presented to show how the basic nanofibrous structure can function as the mid-layer in a three-tier membrane approach. Together with an

Laser Light and X-Ray Scattering: With Evolution of Electrospinning Technology in Biomedical Applications and Water Purification
Benjamin Chu

ISBN 978-981-4968-88-1 (Hardcover), 978-1-003-54074-8 (eBook)
www.jennystanford.com

illustration on the top-layer formation, some general results are being demonstrated, with the main topic for the chapter being biodegradable polymers for biology applications, including key claims. Finally, a statement by President John F. Kennedy in a 1961 press conference, together with a remark made by Mark Twain is used to respond to the question: Is Water our Next Challenge? The need for water purification is presented as a major challenge to the world in the current environment as affected by Covid-19 and climate change on a global scale. Invariably, politics comes into play. Considering the competition between two major nations with the largest GDP, the United States and China, together with other countries, such as sub-Saharan Africa, where countries simply need safe drinking water, the challenge is, indeed, on a global scale.

This chapter has six sections (4.2–4.7). It is intended mainly to provide connections on how the topics of Chapters 2 and 3 can be connected together and for applications to a variety of challenges of current interests, such as filtration of air and water as well as biomedical applications. More specifically, the discussions are being expanded and presented in Sections 4.2 to 4.6 as follows.

4.2. **Thoughts on Air-Gas Filtration:** More precisely, we may consider this subsection as Air Filtration. The use of fibrous membranes for fluids and air filtration processes has fundamental similarities. The details can be very different because of the differences in the nature of gas or fluid properties and the impurities involved. The emphasis in this subsection deals mainly with air filtration.

4.3. **Biomedical Application and Examples:** In addition to purifications for gases or fluids, the examples extend to biological applications, where new frontiers are being expanded, especially because function can often supersede the cost limitations in biomedical applications.

4.4. **Water Purifications with Fibrous Membranes:** The essence of the technique and the importance on water purification are presented.

4.5. **Highly Permeable Nanofibrous Membranes & Luffa Membranes:** Different forms of membranes and oil-absorbent sponges are presented.

4.6. **Structural Characterizations of Cellulose Nanofibers.** Characterization of Cellulose Nanofibers with applications to separation membranes for water purification is presented, partially to emphasize the importance of using plant-based, instead of petroleum-based, materials to resolve challenges on water (and air) purification as well as many problems associated with biomedical applications.

4.7. Aims and Looking Forward.

In the Applications of Polymer, the primary topics that we have discussed on liquid (and air) filtration cover a small section in the panel on Filtration Media. Moreover, even in the filtration media panel, the broader term for *air* becomes *gas* and that for *water* becomes *liquid,* while only molecular filtration was being mentioned. It is important to realize that polymer (and inorganic as well as composite) nanofibers have many uses, thereby making the production of nanofibers by means of electro-spinning/blowing processes a worthy technology.

Applications of Polymer Nanofibers in Water Purification

In the Applications of Polymer Nanofibers, S. Ramakrishna and coworkers [1] presented a very informative figure, depicting the utility of polymer nanofibers over a range of fields, including cosmetic skin masks, protective clothing, tissue engineering, and a broad range of industrial applications. On an even broader scientific scale, nanofibers can be used in the life sciences as drug delivery carriers, in hemostatic devices, and for wound dressing, as well as nanosensors in thermal, piezoelectric, biochemical and chemical probes. Finally, nanofibers have become an essential component in filtration media for liquid, gas and molecular purification. In the current discussion, we deal with water purification where nanofibers play a pivotal role.

4.1.1 Thoughts on Air-Gas Filtration with Nanofibers

In using the fibrous membranes, it is worthwhile to consider the information related to the structure and properties of those membranes formed by nanofibers. As we have started our discussions on water purification, an outline on the water challenge is first presented.

For water purification, we consider the sources of water for purification, e.g., from rivers, lakes, oceans, or discharged water from homes or/and industries. The purification process is a complex issue, as has been discussed briefly in Chapter 3. Now, we limit electro-spinning/blowing processes as a platform technology and broaden the perspective to other potential applications, as has been illustrated in the above discussions on Applications of Polymer Nanofibers.

The schematic drawings of woven and non-woven fabrics in Section 4.1.2 show the difference in structure for the two types of fabrics, with the woven fabrics having a well-defined structure as shown in (a). However, when the fiber diameter goes down to the micron-size range, the non-woven fabrics, as shown in (b), the middle panel, reveal an interesting feature, especially for multilayer structures. To strengthen the non-woven fabric, various methods have been developed to add soldered junctions, as shown in (c).

It should be noted that the clothes we wear normally use woven fabrics. What we wear or use often has a few layers of woven fabrics with spaces in between the fabric layers. If we take a close look at the structures of woven and non-woven fabrics, we see the differences as shown in Section 4.1.2.

4.1.2 Schematic Drawings of Woven and Non-Woven Fabrics

Desirable characteristics of high porosity and inter-connectivity are illustrated in Figure 4.1.

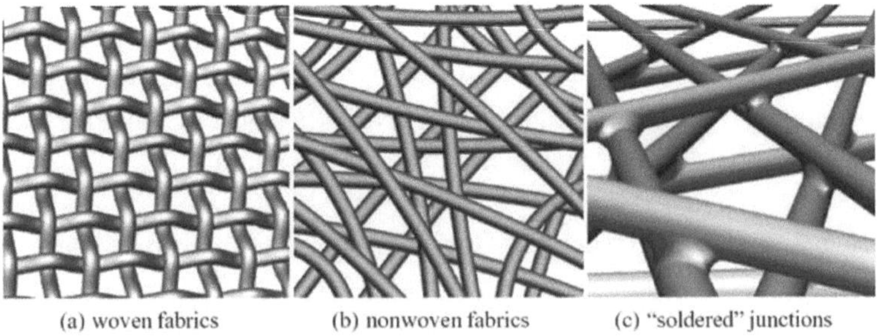

Figure 4.1 Desirable characteristics of high porosity and inter-connectivity.

The integrity of the porous non-woven fabrics can be strengthened further by the addition of "soldered" junctions, as shown in Figure 4.1 (c), with chemical/physical means during the processing stage. It is important for us to know some basic features of a filtration membrane, i.e., the filtration membrane system often consists of three layers, as shown in Figure 4.2.

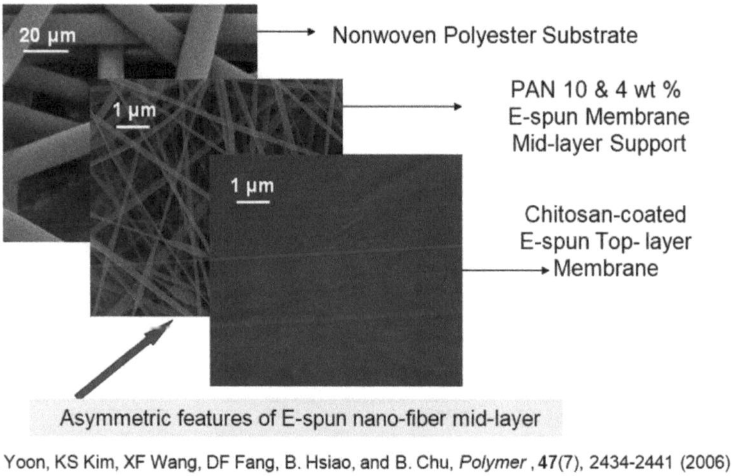

Figure 4.2 Asymmetric features of E-spun nanofiber mid-layer.

SEM images of PAN/Chitosan composite membrane surfaces are presented in Figure 4.2. The bottom layer is a non-woven polyester substrate (with a scale of 20 μm). The mid-layer support is an electro-spun membrane and the top layer forms a very thin membrane to carry out separations of species, with a potential capability that can reach down to the molecular level, though not for the figure being illustrated above.

In considering the "soldered" junctions, we can think of the schematic drawings of woven and non-woven fabrics further down to pore sizes reaching the molecular level, thereby with a separation capability down to molecular sizes. The key factor is the porosity of the separation layer. In this respect,

- the porosity for non-woven membranes is of the order of ~0.7–0.8
- the porosity for micro-trussed layer is ~0.5

While the porosity of both such layers can improve, the cost factor may become an important consideration, i.e., in membrane design, it is always worthwhile to remember *efficiency versus cost.*

For water purification membranes, we show examples of polymers tested by using the three-tier approach, which has been partially described in the previous two figures.

4.1.3 Examples of Polymers Tested

Figure 4.3 depicts the three-tier approach. The polymers used for the coating layer, the mid-layer support and the support, together with electron micrographs of polymer materials used, are presented.

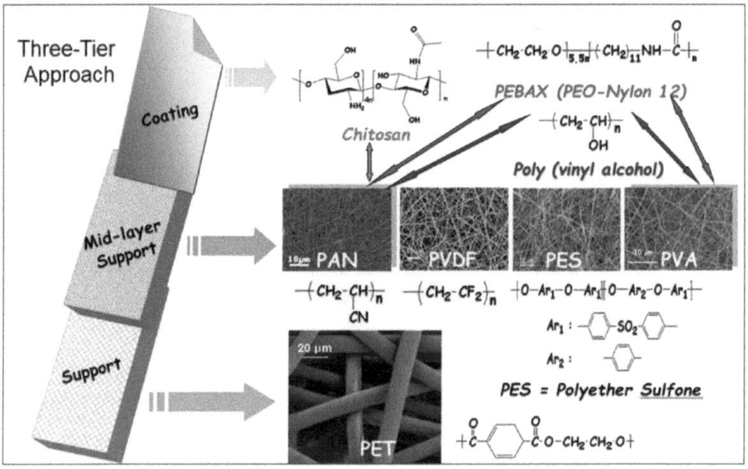

Figure 4.3 The three-tier approach.

Figures 4.4 and 4.5 show PAN E-spun density and diameter control as well as fiber size versus membrane porosity. These are experimental tests used to determine porous membrane fabrication that will be designed to optimize the filtration membrane performance for a specific application. The two processes, as shown, demonstrate some of the steps being used in the separation membrane technology.

In Figure 4.4, electron micrographs of PAN E-spun fibers from 4, 6, 8, and 10 wt% PAN solutions show, respectively, average fiber diameter sizes of 124, 280, 682, and 720 nm. Together with varying conditions of the electro-spinning process, the E-spun PAN nanofiber diameters can be controlled to vary by a factor of about 6, in this example.

In Figure 4.5, electron micrographs of PAN E-spun fibers from 6, 8, and 12 wt% PAN solutions show, respectively, fiber diameter distributions of PAN E-spun fibers. Porosity (%) and average fiber diameter (nm) of the three samples are shown in the lower left plot. The interesting result tells us that *while we have increased the fiber diameter by **a factor of >8**, the porosity has increased by only about **3%**.*

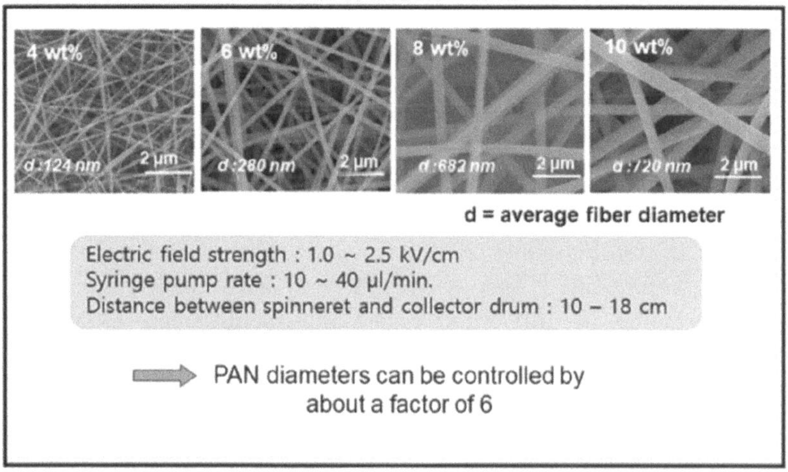

Figure 4.4 PAN E-spun fiber density and diameter control.

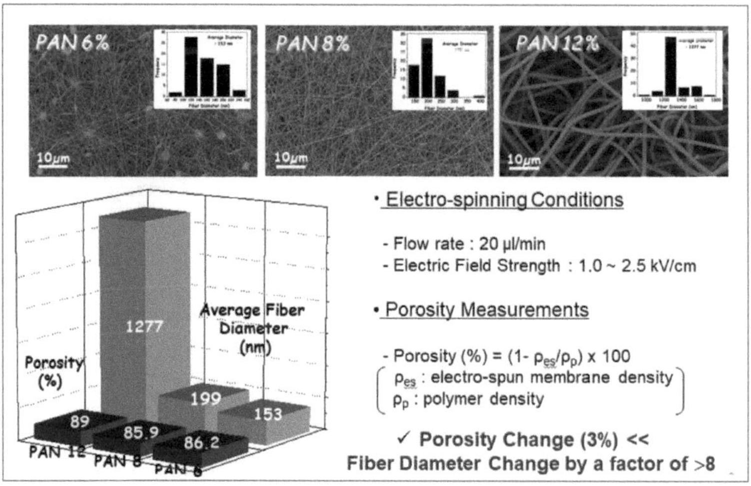

Figure 4.5 PAN E-spun fiber sizes and membrane porosities.

Figures 4.4 and 4.5 have demonstrated the importance in the fiber diameter used versus the corresponding porosity change of the layer. The reader should remember this information when dealing with filtration membranes.

4.1.3.1 A qualitative argument

In order to explain the importance of high porosity so as to achieve a high flow rate for the filtration process, a qualitative argument is presented as follows:

1. **E-Spun Fiber Sizes and Porosities**

   In Figure 4.6, the E-spun membrane, as shown by a schematic diagram with 80% void and 20% materials, is laterally being compressed to a solid 20% material, leaving an 80% void volume. Then, in an incoming water flow, as shown in the mid-panel, 20% volume is impermeable and 80% is permeable. Thus, if we increase the thickness of the membrane, the water flow rate should remain essentially the same, as shown in the right panel.

   We then carried out experiments to prove the illustration as shown in Figure 4.6.

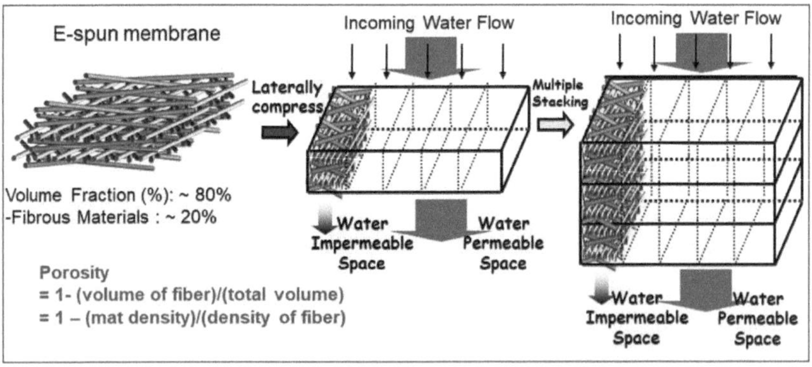

Figure 4.6 E-spun fiber sizes and porosities with virtually no effect of fiber diameter on porosity. Water essentially flows only into empty space, implying that there is a minimum flux reduction by a membrane thickness increase.

2. **Hydraulic Resistance of PAN E-Spun Membranes**

   Based on the test results of Figures 4.6 and 4.7, we have illustrated that the use of *non-woven nanofibrous membranes* is one viable pathway for a key component that should be taken into account in the development of water purification systems.

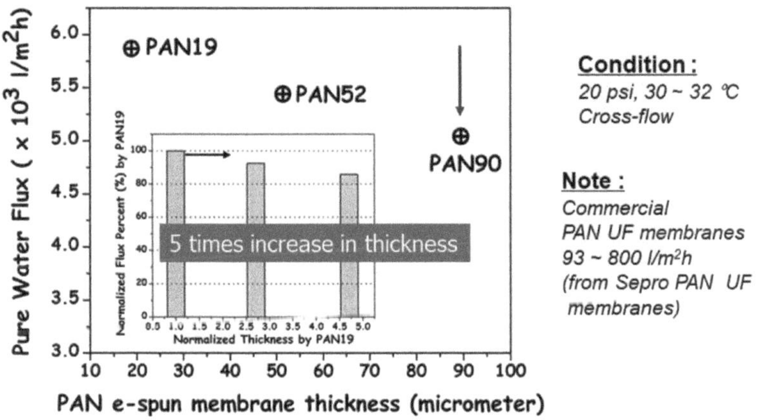

Figure 4.7 Hydraulic resistance of PAN E-spun membranes. Low hydraulic resistance: Only 14% flux decrease by a factor of 5 in the thickness increase. Pure water flux for PAN E-spun membrane (5~6 × 10^3 l/m^{2}h) is ~10 times higher than commercial PAN UF membranes (100~800 l/m^{2}h).

On close examination, the test has covered one idealized aspect of E-spun membranes. The overall process of formulating a filtration membrane involves a broad range of pathways because of the large number of interacting variables. For example, in depositing ultrathin polyamide (PA) coating on PAN electro-spun membrane, an interfacial polymerization process, as shown in Figure 4.8, is involved. Thus, the reader should try to remember the complex interactions among the many variables associated with the process and to work out a scheme so as to try to minimize the steps needed in the formulation in order to achieve the intended result. SEM images on the membrane structures with two different types of support, as shown in Figure 4.9, can be a good guideline to see what has been fabricated. The product can then be viewed in comparison with a known commercial product (or different conditions of preparations). We can see the need for a carefully designed plan that can be modified according to the performance of the overall system.

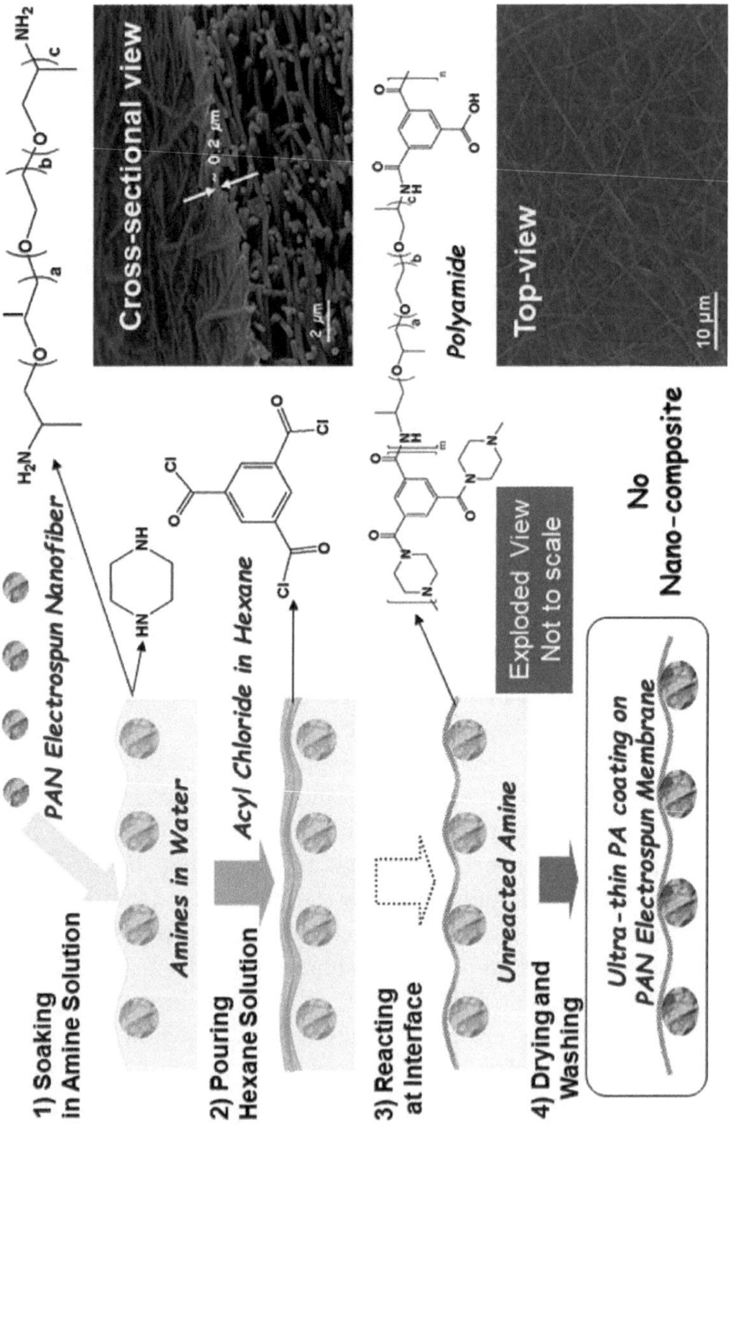

Figure 4.8 Interfacial polymerization.

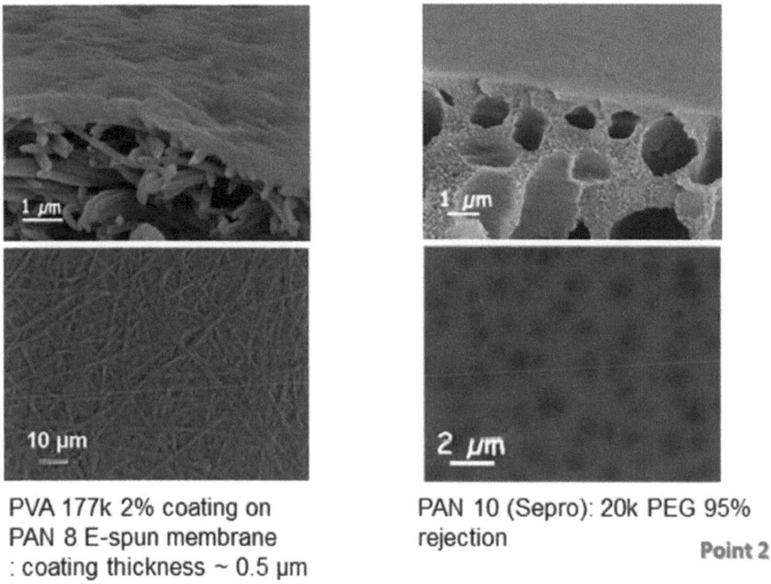

Figure 4.9 SEM images of the membrane structures: PVA versus PAN.

In comparison with different membrane structures, for example, for a composite membrane versus a commercial membrane, both with a PVA hydrogel coating layer, the effective hydraulic resistance layer will behave differently, as shown in Figure 4.10. Figure 4.11 reveals that the difference in PAN E-span membrane and a PAN UF membrane illustrates the difference in the effective hydraulic resistance layer for water flowing through the two different PAN membranes.

The PVA coating layer thickness dependence on pure water flux by using different PVA coating solution concentrations shows that the effect of supporting layer has a large effect on the pure water flux. Figure 4.12 shows emulsified oily water filtration for PVA-PAN over a time period of 190 hours with the E-spun membrane having a flux of up to 12 times higher when compared with commercial PAN10 (Sepro).

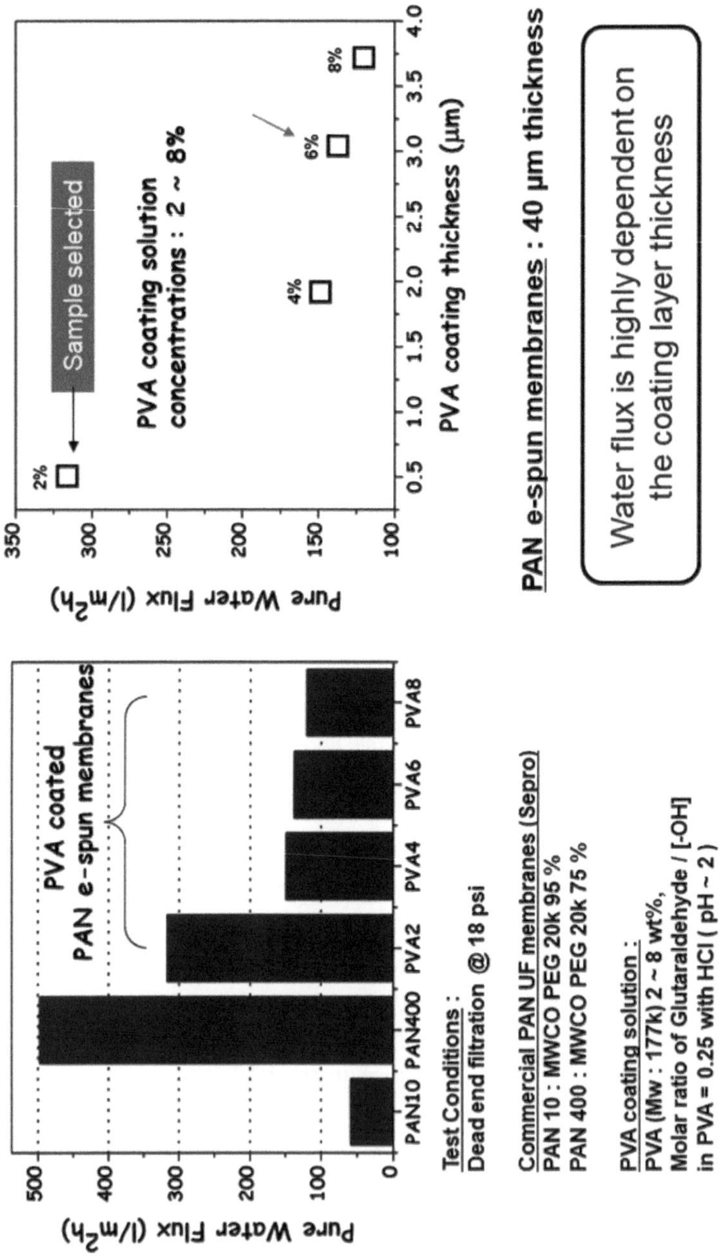

Figure 4.10 PVA coating layer dependence on pure water flux behavior.

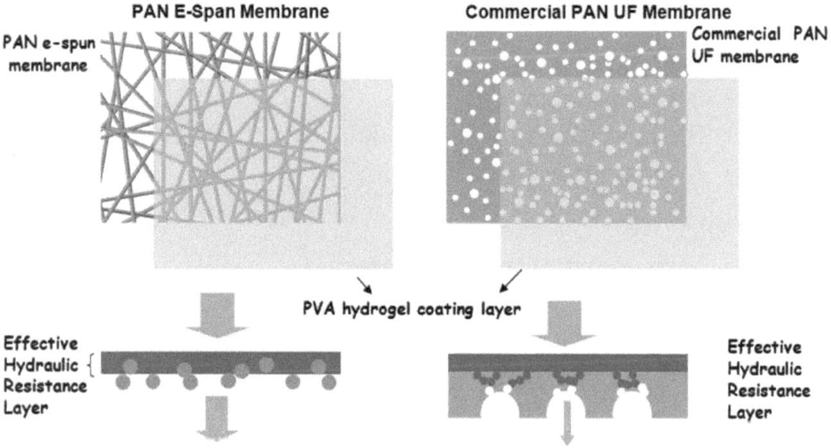

Figure 4.11 Different membrane structures: Composite E-spun membrane vs. commercial membrane.

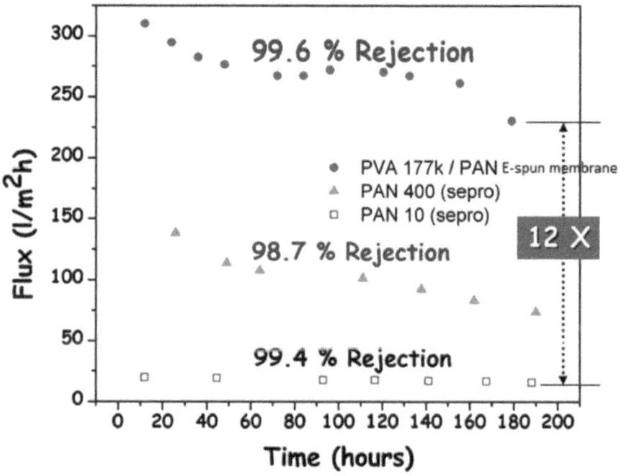

Figure 4.12 Emulsified oily water filtration for PVA-PAN for 190 hours. Test conditions: Cross-flow mode, inlet pressure—90 psi, 30–32 °C; feed—soybean oil (1350 ppm) + DC 193 fluid (150 ppm). Upper (99.6% Rejection), middle (98.7% Rejection) and Lower (99.4% Rejection) curves refer to PVA 177K / PAN, PAN (sepro), and PAN 10 (sepro), respectively.

4.1.3.2 Example

Scaffolds formed by copolymers of lactide and glycolide (PLGA) via electro-spinning have been developed for controlled DNA delivery with the advantages and disadvantages being described as follows:

Advantages (in scaffolds)

- Degradability
- Biocompatibility
- High surface area/volume ratio
- Adjustable mechanical property

Disadvantages

- Limited control on DNA release rate
- Degradation of DNA molecules after release
- No transfection activity of DNA detected

Adding $PLA_{10}PEG_{78}PLA_{10}$ tri-block copolymers to the scaffold is an example provided as shown. Here we try to provide an example to show how *tri-block copolymers* interact with *DNA* under electro-spinning conditions with a solution containing 94% DMF + 6% water. The key conditions are presented in Figure 4.13.

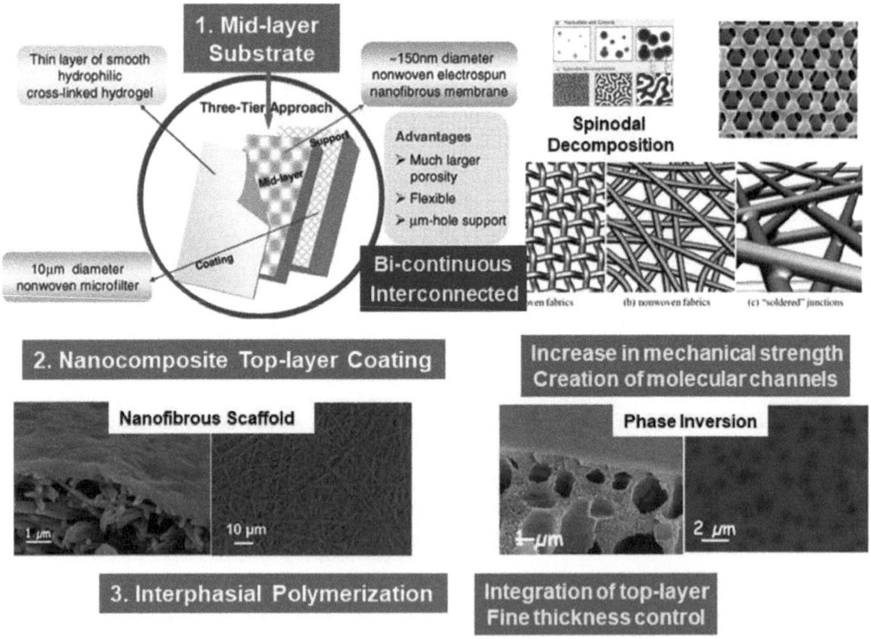

Figure 4.13 Key conditions of mid-layer substrate, nanocomposite top-layer coating, and interphasial polymerization for controlled DNA delivery. The bi-continuous interconnected fibrous format on soldering-like attachments are amplified in Figure 4.37.

We end with a statement by President John F. Kennedy in a 1961 press conference:

> "We have made some exceptional scientific advances in the last decade, and some of them are not as spectacular as the man-in-space, or as the first sputnik, but they are important. I have said that I thought that if we could ever competitively, at a cheap rate, get fresh water from salt water, that it would be in the long-range interests of humanity which would really dwarf any other scientific accomplishments."
>
> http://www.presidency.ucsb.edu/ws/index.php?pid=8055

Paraphrasing a remark made by Mark Twain:

> *Sophisticated research topic is like wine. It needs to be savored and appreciated.*
> *My topic is like water. But everyone drinks water.*

Our emphasis is on themes dealing with common daily life and of larger impacts to society:

We need to develop an economy that preserves and respects the environment.

Is Water Our Next Challenge?

We try to impose the question on how important "water" is for the well-being of our society. Predictions seem to suggest the following argument, i.e., if we take into account of the following conditions:

- By doubling of human population every 40 years
- The prediction of Global warming
- Water as a finite resource

Then, water is, indeed, a fundamental need to the well-being of our society.

What is a **Possible Solution** to the above **Challenge**? We should consider the following environment and conditions that may offer a viable solution to the above challenge, i.e., we should think of

- developing our technology to be often linked with politics and the environment.
- achieving current improvements by a factor of 10–100—as a target for the water challenge.
- changing our lifestyle.

4.2 Thoughts on Air/Gas Filtration

Air/gas and water/oil purifications, by means of filtration membranes, follow similar physical schemes. In the current context, Chapter 3 discussed the key components for a separation membrane whereby the electro-spinning/blowing technology can produce nanofibrous membranes, having a better mid-layer support for the fluid separation membranes. In air/gas purifications, such as for face/gas masks, high permeability often becomes an essential requirement. Thus, the main concern deals with *breathability* in the design. The fundamental non-woven structure coming from electro-spinning/blowing technology has remained a very competitive approach for air/gas filtration. The primary concerns are related to the production of such porous structures, in combination with other schemes, such as chemical modifications on the surface of nanoporous structures, or/and introduction of aggregated polymer chains or globules to assist or deter flow of air or specific gaseous species. Then, the filtration of air/gas molecules begins to provide different emphasis in the design and optimization of such processes.

It is noted that, in the air/gas filtration, the scope has been expanded. While one thinks of air with main ingredients of oxygen, nitrogen, and carbon dioxide, gas can be materials in the gaseous state, or small particles in solid or/and liquid forms in the air. Thus, the purification process can include passage of some species and retention of others. In addition to the physics used in the separation process, there is chemistry associated with the nature of the species involved. It is essential to take into account of both physics and chemistry to design and to optimize each specific separation process.

4.2.1 Brief Outline for Section 4.2

In the air/gas filtration (Section 4.2), it is clear that living beings cannot live long without air and people need clean air to have a reasonable quality of life. Some filtering mechanisms and information on particle size versus fiber diameter are presented. As basic knowledge is needed to consider how face masks work, the fundamental equations, together with motivation and approach are being included in the discussion. Advantages for separations at a molecular level by using structures derived from the *spinodal decomposition* process are also being briefly considered especially because of future applications for filtrations of species on a molecular scale.

Figure 4.14 An example of air quality in Beijing in October 2013.

4.2.2 Air Quality to Suffer with Global Warming

There are many different sources of pollution. For example, aside from industrial and agricultural sources, as well as climate change, some causes of poor air quality have been around for a very long time, such as on bad days in Beijing with (fine) sand storms all the way coming from the north. While the government can provide pathways to improve air quality, the use of face masks is here to stay for years to come in the foreseeable future. In addition to the face masks, there are larger capacity air filters for the purification of air in rooms and assembly halls, such as theaters. Thus, the need for further improvements on air filtration under a variety of environment also exists.

It should be noted that the current approach, under discussion, of using ultrafine (primarily petroleum-based polymeric) fibers, being fabricated by the electro-spinning/-blowing technology, to formulate a critical component of the filter, represents *one* promising pathway. There are other approaches which can be used to solve specific needs. Thus, the reader should be aware of the role the current approach can play in the filtration field.

For face masks, a brief discussion of some filtration mechanisms, as it was presented by the *New York Times* on April 27, 2003, is being summarized. Here, early references are being used to show the reader that the challenges have been around for some time.

The air filtration process can be achieved by primarily considering the following:

1. Electrostatic interactions due to charges of airborne particles
2. Sedimentation of heavy particles
3. Diffusion of lighter particles

The relationship between the fiber diameter and particle size is intriguing. It shows the scales of particle size range (roughly from 10 nm to 1 μm) versus a couple of (e.g., polyethylene and polypropylene) fibers. A gallery of particles of different sizes and shapes, starting with the Rhinovirus at 0.018 μm, *Mycobacterium tuberculosis* at 0.2 μm but now with a cylindrical shape, *Neisseria meningitides* at 0.6 μm in spherical shape, etc. Thus, we know that for air filtration, we need to know the nature and amount of impurities that we are dealing with.

4.2.3 Some Filtering Mechanisms for Face Masks

It is interesting to note that when we sneeze, there are mostly ~6 micron-size particles and one sneeze gives out about 350,000 such particles, including charged particles, heavy and light particles for the face mask to screen out. Thus, new face masks invariably use a range of fine-diameter fibers, often with fibers having surface active properties, to screen out those particles, as well as undesirable finer particles or even vapors.

There are three primary mechanisms for face masks:

1. *Electrostatic interactions* due to charge of airborne particles.
2. *Sedimentation* of heavy particles.
3. *Diffusion*: small, lighter particles can bounce randomly in the airstream, increasing the probability of particles to make contact with a fiber or fibers of the face mask, and being trapped.

4.2.4 Fiber Diameter versus Particle Size

While the particle size of interest can range from about 10 nm to micron sizes, the fiber diameter used in the face masks can be of the order of microns or larger. It is the spacing between fibers that can cover a broad size range, depending partly on the primary interest in the face mask design. There is a gallery of particles (virus, bacteria, etc.), such as Rhinovirus at 0.018 micrometer, influenza at 0.08 micrometer, and *Neisseria meningitides* at 0.6 micrometer, that one may wish to screen out. Thus, the optimal design of face masks depends partially on specific targets.

4.2.5 Motivation and Approach

To broaden the scope from air filtration to gas filtration and further extending the scope to filtration at a (much smaller) molecular level, the requirements mean that the separation technology should be expanded toward a broader perspective to include, for examples, viruses, bacteria, pollen, and pollutants. Based on this broadened scope, we can consider an interdisciplinary approach in terms of the following sub-topics:

- Design
- Synthesis
- Processing
- Characterization
- Testing

However, it is even more important to consider the initial requirements in the design of an air filter, starting with Gas Permeation Selectivity (F), a key expression for evaluating face mask performance:

$$F = PA\,\Delta p/\sigma,$$

where F is the volumetric flow rate (or the gaseous flux), σ the membrane thickness, P the intrinsic permeability coefficient that characterizes the specific membrane permeation (where P/σ is the normalized flux), and Δp, the applied pressure difference.

F is linearly proportional to *P, A, and* Δp but is inversely proportional to the membrane thickness (σ). Then, *thinner membranes* can be more efficient. However, a balance with other requirements is essential.

For separations at sizes reaching the molecular level, we think of porous structures with a much smaller size, such as structures which one encounters in the spinnodal decomposition of polymer blend/solution, as shown in Figure 4.15.

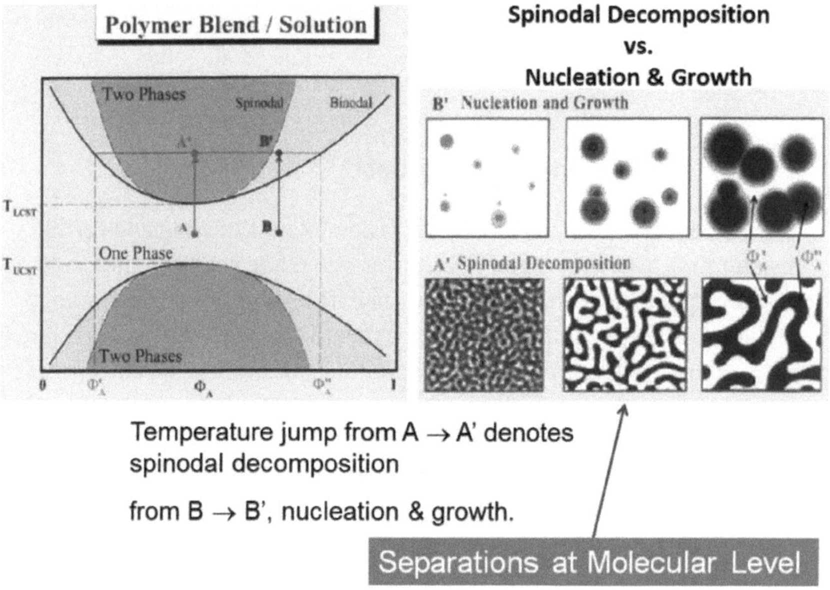

Figure 4.15 Schematic phase diagram of polymer blend/solution.

4.2.6 Advantages for Separations at a Molecular Level

1. The amount of pores (or the dissolved component) is adjustable.
2. The correlation length (pore size) is adjustable.
3. The characteristic surface is based on the nature of the polymer used. In other words, we can take advantage of chemical modifications.
4. The bi-continuous nature of the pores implies that blockage should be less of a problem, due to the connectivity of the

pores, as shown schematically for the system undergoing the spinodal decomposition.

Without going further on the details for air filtration at this time, we think of the important considerations as follows.

4.2.7 Further Considerations

Air filtration can be expanded to include gas filtration and separations at a molecular level. Then, large-scale and more complex productions may come into play. Thus, we need to pay attention to

1. Cost: Success at a laboratory level is insufficient. It is important to consider the scale-up route early in the process.
2. Scale-up: As indicated in cost considerations, research in laboratory is only a fraction of the cost of large-scale productions.
3. Pore size ranges: Concern of pore size ranges is related to materials properties and processing procedures.
4. Porosity limit: Same as pore size ranges. Materials and process changes for large-scale operations should be considered early in the process.

4.3 Biomedical Application and Illustrations Based on Chapters 2 and 3

In the evolution of electro-spinning technology, while the discussions are concerned mainly with filtration processes by taking advantage of nanofibers that can be fabricated by means of electro-spinning/-blowing technology, it is worthwhile to note that nanofibers, coupling with other fabrication methods, have a *broader range* of very useful applications, in the fields of biology and other engineering disciplines, in addition to water and air filtrations. For this reason, a brief discussion is presented below. As an example, for biomedical applications, useful functions for medical applications can often be more important than cost justification. Thus, the final aim for biomedical applications may have a more relaxed priority, i.e., function may supersede cost. This section starts with the advantages and disadvantages of polymers for biomedical applications.

4.3.1 Overview

Several examples of polymers for biomedical applications, including the use of copolymers of lactide and glycolide (PLGA), because of their desirable properties on biodegradability and biocompatibility, are presented. Usually, each polymer example has advantages and disadvantages in connection with specific objectives in biomedical applications. Thus, the selection for presentation depends on a range of factors, based primarily on specific target(s) for the application. The examples now being shown, such as biodegradable polyesters and polysaccharides, have a broad range of applications. One example system on DNA-CTAB complex formation, leading to macro-lattice structures, coil-globule transition, and biodegradable electro-spun scaffolds containing PLGA (90% PLGA, 10% BC (block copolymer))/b-gal plasmid DNA has now *jumped* to *a more advanced level of discussion*. Such a scaffold has the suitable morphology that will enable endogenous cells to migrate into it, adhere, proliferate, and initiate new matrix/osteoid production. Adhesion and proliferation of MC3T3 cells were determined by the incorporation of a dye. Verification of Integrity of released DNA from a 30% PLGA/BC (block copolymer) scaffold was carried out by using gel electro-phoresis, with final results being summarized. The specific illustration provides an interesting example whereby the two techniques, as discussed in Chapter 2 (laser light scattering and synchrotron X-ray scattering) and Chapter 3 (electro-spun membranes for filtration and biomedical applications) are involved and the processes are also connected with other related filtration methods, in order to achieve the final goals in an application process. The pathway shall likely be similar for future development of new systems. The current presentation sets an example of such processes. It is noted that in applications, especially for the biomedical field, the techniques of the current interest in the book, Chapter 2 (laser light scattering, synchrotron X-ray scattering) and Chapter 3 (electro-spinning/

-blowing technology), are only two among many of the physical techniques. Examples are given to show how those two techniques can be combined to elucidate the structures of interest. The reader should not overlook the presence of other physical techniques, such as electrophoresis, NMR, different forms of microscopy, just to name a few.

4.3.2 Polymers for Biomedical Applications

Scaffolds formed by copolymers of lactide and glycolide (PLGA) via electro-spinning have been developed for controlled DNA delivery.

Advantages (in scaffolds)

- Degradability
- Biocompatibility
- High surface area/volume ratio
- Adjustable mechanical property

Disadvantages (so far)

- Limited control on DNA release rate
- Degradation of DNA molecules after release
- No transfection activity of DNA detected

4.3.3 Advantages and Disadvantages of Polymers Used in Biomedical Applications

Consider *Materials* often used in *Biomedical Applications.* Pay attention that each material has its own associated biological, chemical, and physical properties. Furthermore, mixing of materials will have associated interactions affecting their overall properties. Examples of polymers for biomedical applications are listed in Figure 4.16 for your reference.

The biodegradable polyesters and polysaccharides are polymers often used in biomedical applications. They are being listed for the convenience of readers, especially for those who are less familiar with polymeric materials.

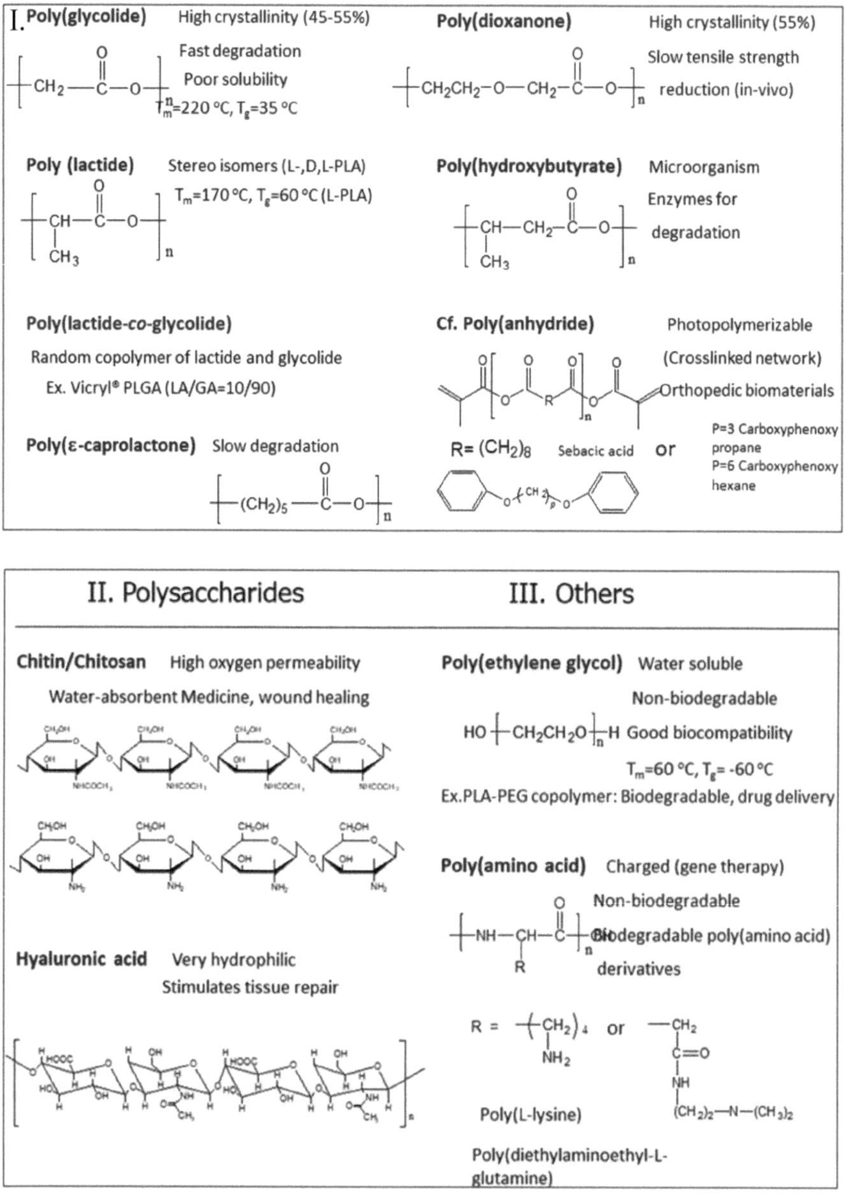

Figure 4.16 Examples of polymers for biomedical applications.

In this subsection, the examples are being used to illustrate a few systems that we have encountered for applications to biomedical applications. It is intended to provide the reader a sense of the complexity that often occurs in such applications. Again, it pays to have a plan with alternative pathways because on the one hand, we have a specific target (or targets), while on the other hand, there are alternative pathways, dependent on the outcome of intermediate investigations, to reach the goal.

An illustration of the systems studied was published by Burger et al. [3]. There were three complexes in equilibrium with a dilute surfactant aqueous solution. The structures can be studied by using a range of instruments, including X-ray scattering. The details have been reported in the reference as shown.

The following two examples deal with more straightforward simpler characterizations by using X-ray diffraction and laser light scattering:

1. X-ray diffraction pattern, unit cell model and P(MAA-co-NIPAM)-CnTAB complexes (Figure 4.17)
2. Coil-to-globule transition of single PNIPAM chains

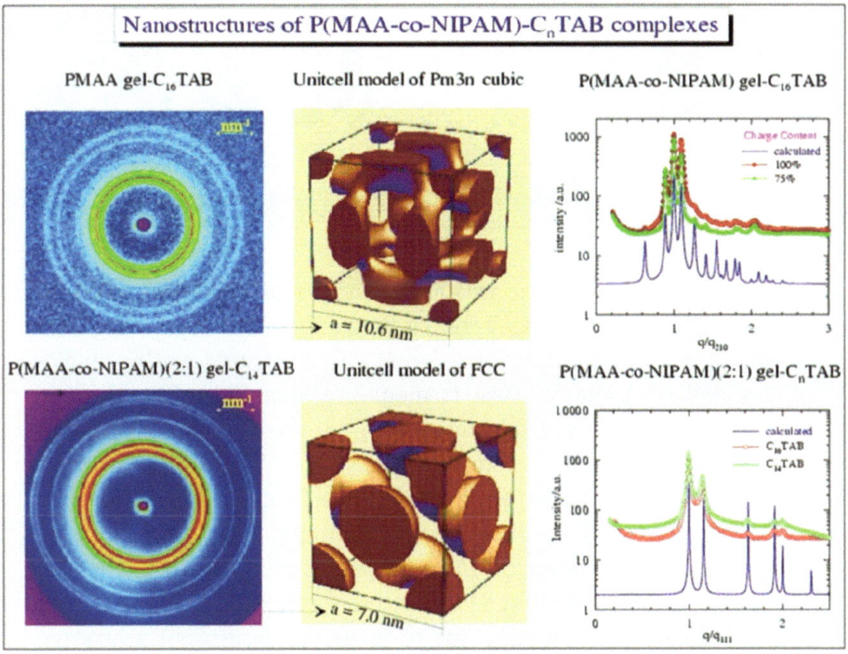

Figure 4.17 X-ray diffraction pattern, unit cell model and P(MAA-co-NIPAM)-CnTAB complexes. If the reader finds difficulty to reach the information for the figure above from a combination of X-ray diffraction, unit cell modeling, and intensity versus q ratio plot, it will be worthwhile to consult the texts on those topics.

Based on the small angle X-ray scattering (SAXS) studies, we can determine the type of complexes that have been formed in terms of macro-lattices. Figure 4.18 shows the peak position ratios of a range of those systems.

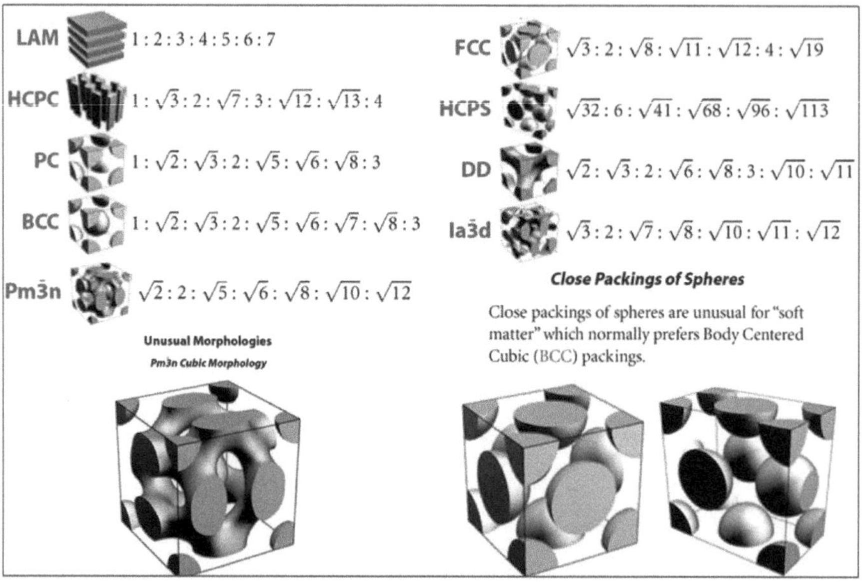

Figure 4.18 Small angle X-ray scattering (SAXS) from macro-lattices (peak position ratios) [3].

Figure 4.19 shows another example of nanostructures of DNA-CTAB complexes. When the double-stranded DNA was denatured in a solution of 0.5 mM EDTA+10 mM NaOH, single-stranded DNA chains were formed. The X-ray scattering pattern of double-stranded DNA, as shown in the left panel, had an inverted hexagonal structure, while that of the single-stranded DNA, as shown in the right panel, had a Pm3n cubic structure.

It is noted that while double-stranded DNA chains can be fairly stiff, single-stranded DNA chains are more flexible, whereby the structural elements can become more complex.

In Chapter 2, the aim was to introduce laser light scattering (LLS) with the use of crossed incident laser beams and then to adapt such a setup in combination with synchrotron X-ray scattering. It was indeed an ambitious undertaking. In the process, BC has

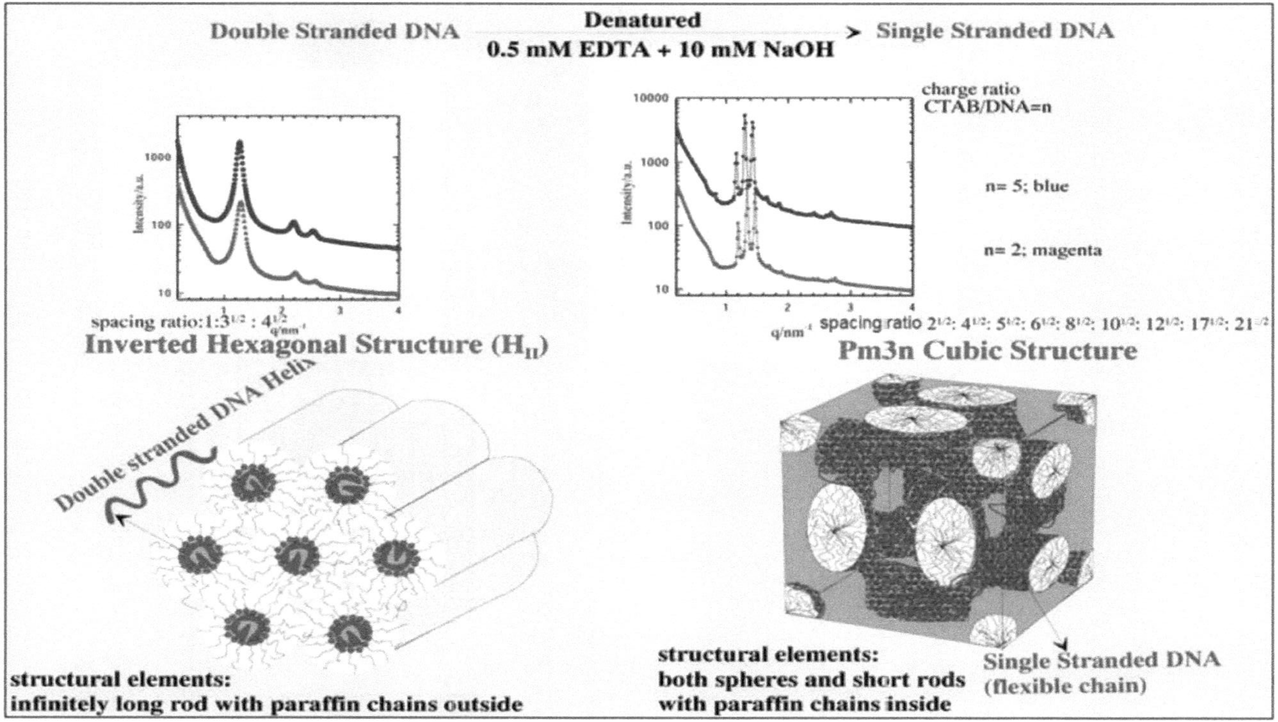

Figure 4.19 Nanostructures of DNA-CTAB complexes. On grayscale and for the upper two figures, the upper and lower curves are, respectively, for n = 5 and n = 2.

not provided sufficient information for the uninitiated reader on the use of single incident beam laser light scattering which has now become readily available from a range of commercial sources, including Brookhaven Instruments, started by two former post-doctorates of BC at Stony Brook.

For the application examples, the author now plans to introduce a couple of light scattering studies to demonstrate that good information can be obtained to resolve very challenging questions. The examples include long-standing problems on:

1. Coil-to-globule translation
2. Behavior of plasmid DNA

We now show the schematics of a laser light scattering spectrometer with associated instrumentation and key information commonly achievable in such an instrument (Figure 4.20).

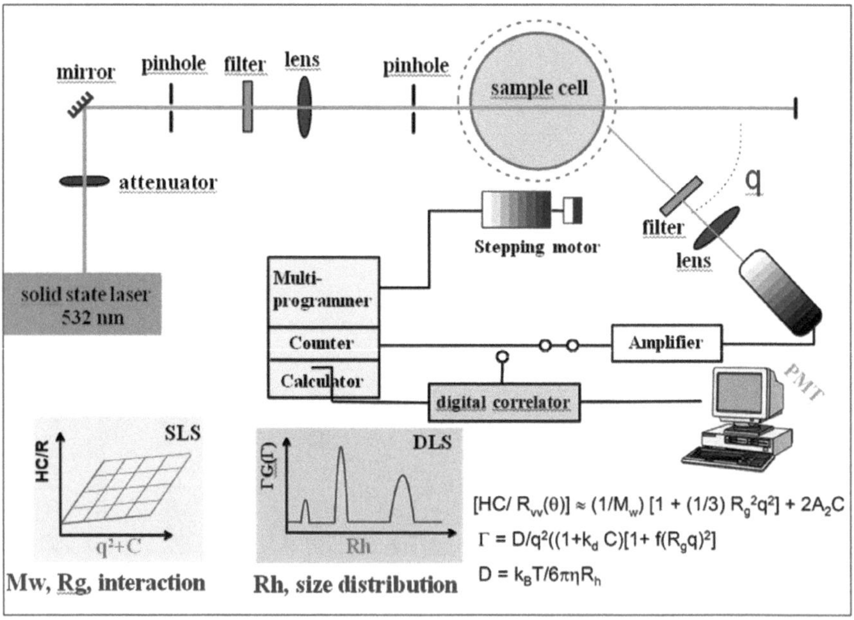

Figure 4.20 Laser light scattering spectrometer with associated obtainable information.

The equations for LLS have been presented in Chapter 2. The primary information from both static and dynamic light scattering has been summarized in Figure 4.20. For static light scattering (SLS) with a Zimm plot, we get the weight average molecular weight (M_W) and the radius of gyration (R_g). For dynamic light scattering (DLS), we get the hydrodynamic radius (R_h).

The light source can be a range of different lasers, including He-Ne, Argon ion, or other forms of diode lasers with appropriate photo-detector for the incoming radiation.

The coil-to-globule transition represents a very challenging study on the collapse of a polymer coil to a globule state. There is a range of precautionary measures to carry out such a study. For example, we need to use a polymer sample with a narrow molecular weight distribution and to carry out the study at very low polymer solution concentrations in order to avoid interactions among polymer molecules. At the same time, the prepared dilute polymer solution should have, in essence, no dust particles. Figure 4.21 illustrates such a study.

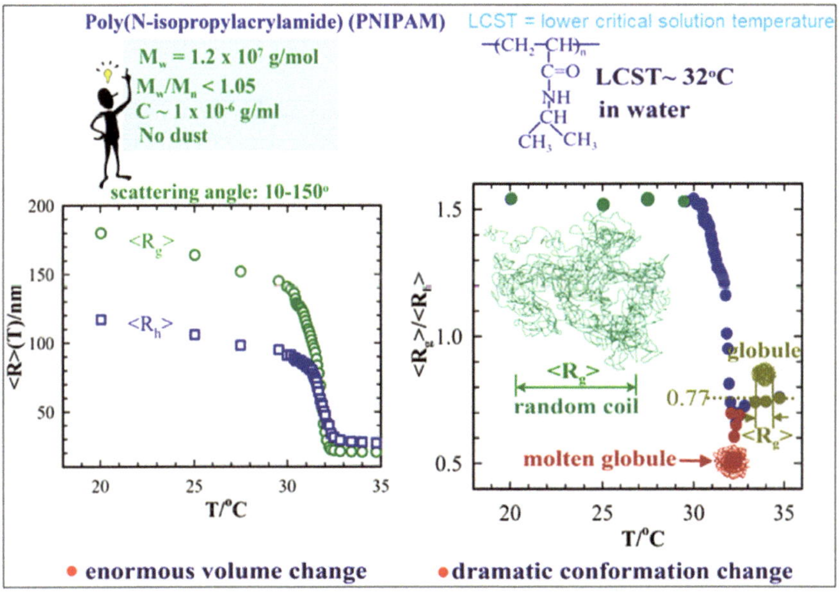

Figure 4.21 Coil-to-globule transition of single PNIPAM chains.

It should be noted that the presence of dust particles will produce a large burst of light scattering signals. Yet if the polymer solution concentrations were too low, another contribution on the entrance and exit of polymer chains in the scattering volume will produce number fluctuations.

Figure 4.21 (left) shows that the experiments were carried out over a large range of scattering angles (θ covering 10° to 150°, beyond the normal range of many commercial light scattering instruments) and the measured radius of gyration (R_g) changed by a factor of ~20, following the temperature change.

Figure 4.21 (right) shows the sketches of the enormous size change in R_g from a random coil ($<R_g>$ in green or grey color) to a globule (in brown/greenish color) and a molten globule ($<R_g>$ in red color). Note that the plot is the ratio of the radius of gyration to the hydrodynamic radius ($<R_g>/<R_h>$) versus temperature. It is noted that with the grey-scale presentation, different colors appear as different shades of grey.

The next example deals with the behavior of plasmid DNA (Figure 4.22). We want to demonstrate how a bio-macromolecule, such as DNA, can be investigated by using SLS and DLS. More importantly, the conformation of the DNA chains changes in different solvent environment.

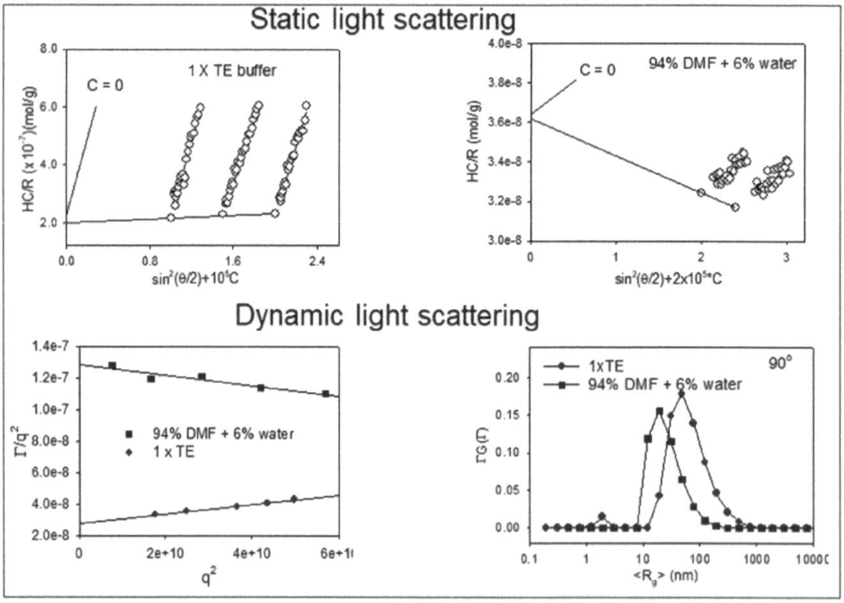

Figure 4.22 Behavior of plasmid DNA. Plasmid DNA showed different LLS behavior in 1× TE buffer and 94% DMF + 6% H_2O.

With static and dynamic light scattering, plasmid DNA in 1× TE buffer and in 94% DMF + 6% water shows different behavior. Furthermore, the interactions of DNA with other molecules can be investigated by looking at the changes in the parameters obtained by a combination of SLS and DLS.

We can view the **physicochemical properties of DNA** in 1× TE buffer and 94% DMF + 6% water in terms of weight average molecular weight (M_w), radius of gyration (R_g), the second virial coefficient (A_2), the diffusion coefficient extrapolated to zero concentration (D_0), the hydrodynamic radius (R_h), and the ratio of R_g/R_h (Figure 4.23). The R_g/R_h ratio tells us that the plasmid DNA forms linear chains in 1× TE buffer and transforms toward a globule form in 94% DMF + 6% water, as the limiting value for the ratio of R_g/R_h of a sphere is 0.775. It is interesting to note that the increase in "molecular weight" from about 5 million to about 45 million suggests an aggregation and condensation of the plasmid DNA.

	1 x TE	94%DMF
M_w /10^6 g-mol^{-1}	***4.81 ± 0.21***	***44.6 ± 5.2***
R_g (nm)	***135 ± 6***	***16.6 ± 2.6***
A_2 (mol cm^3/g^2)	8.0 x 10^{-4}	-2.0 x10^{-4}
D_0 / 10^{-8}cm^2-s^{-1}	2.78 ± 0.13	12.4 ± 1.1
R_h (nm)	***87 ± 2***	***17.2 ± 2.0***
R_g/R_h	***1.55 ± 0.09*** **(>1.5 for linear chains)**	***0.97 ± 0.15*** (0.775 for solid spheres)

94% DMF + 6% water is a poor solvent for DNA.

Figure 4.23 Physicochemical properties of DNA.

The shrinkage in size and the increase in molecular weight suggested that DNA molecules were condensed and aggregated.

DNA started to precipitate when the concentration reached 2 × 10^{-5} g/ml.

Figure 4.24 shows the electron micrographs of coil-to-globule transition of $L_{10}E_{78}L_{10}$ in TE buffer and 94% DMF + 6% H_2O.

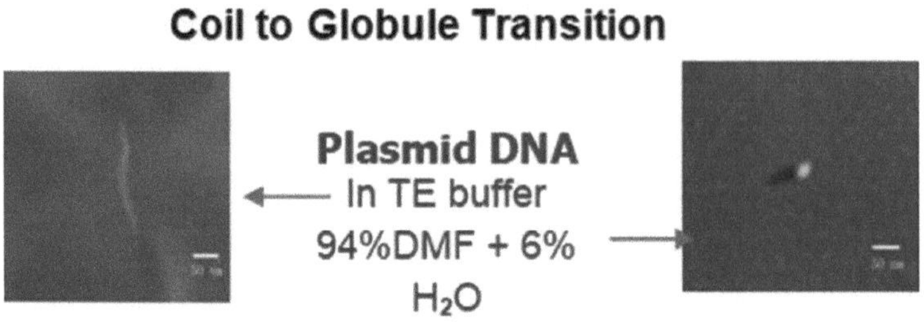

Figure 4.24 Coil to globule transition of plasmid DNA.

We utilized electro-spinning to fabricate synthetic polymer/DNA composite scaffolds for therapeutic applications in gene delivery for tissue engineering. The scaffolds had non-woven nanofiber membrane structures, composed predominantly of poly(lactide-co-glycolide) (PLGA) random copolymers and a poly (d,l-lactide)–poly(ethylene glycol) (PLA–PEG) block copolymer. In the initial study (Figure 4.25 and Table 4.1), the interactions of plasmid DNA with a tri-block copolymer consisting of a polyethylene glycol mid-block (E) and two end blocks of polylactide (L) were carried out.

- In 94% DMF + 6% water, LEL showed a molecular weight (181K), much larger than that obtained from GPC (5.9K) and ^{1}HNMR (4.6K), suggesting the formation of aggregates.
- The value of R_g/R_h, which was much less than 1.5, indicated that LEL was not in the random coil conformation.

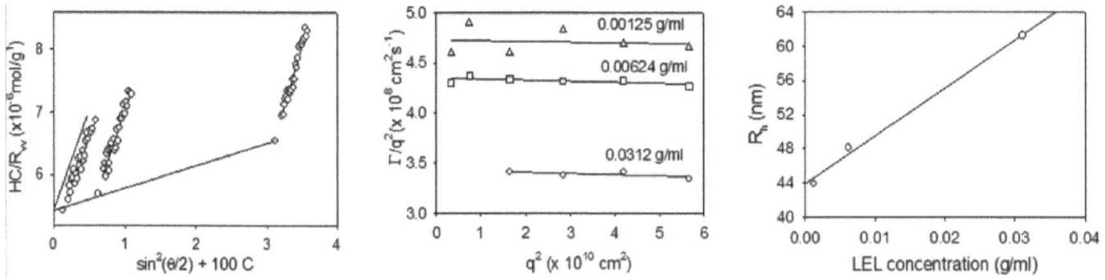

M_w (mol/g)	R_g (nm)	A_2 / 10^{-4} mol cm^3-g^{-2}	$D_o/10^{-8}$ cm^2-s^{-1}	R_h (nm)	R_g/R_h
181 ± 7K	39 ± 2	2.6 ± 0.1	4.7 ± 0.2	44 ± 2	0.89

Figure 4.25a Laser light scattering of $L_{10}E_{78}L_{10}$.

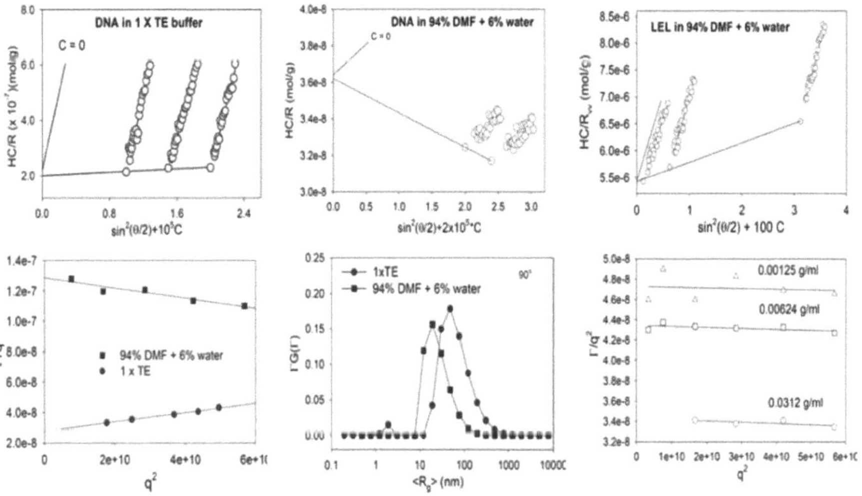

Figure 4.25b Laser light scattering on DNA and $L_{10}E_{78}L_{10}$.

Some earlier LLS results on DNA in 1× TE buffer and in 94% DMF + 6% water have been included to have a comparison with the LLS results on $L_{10}E_{78}L_{10}$ in 94% DMF + 6% water, as listed in Table 4.1. In the lower middle panel of Figure 4.26, the size distribution was also presented. Physical-chemical properties of mixtures of DNA and $L_{10}E_{78}L_{10}$ are listed in Table 4.2.

Table 4.1 Physicochemical properties of plasmid DNA and LEL

	DNA in 1x TE buffer	DNA in 94% DMF + 6 % water	LEL in 94% DMF + 6% water
M_w (10^6 g/mol)	4.81 ± 0.21	43.1 ± 4.2	0.181 ± 0.007
R_g (nm)	135 ± 6	15.2 ± 1.8	39 ± 2
A_2 (Mol cm³ g⁻²)	8.0 × 10⁻⁴	-3.7 × 10⁻⁴	2.6 × 10⁻⁴
D_0 (10^{-8} cm² s⁻¹)	2.78 ± 0.13	12.4 ± 1.1	4.7 ± 0.2
R_h (nm)	87 ± 2	17.2 ± 2.0	44 ± 2
R_g/R_h	1.55 ± 0.09	0.88 ± 0.15	0.89 ± 0.04

In 94% + 6% water:

❑ Conformation of DNA was changed from random coil to globule due to poor solvent quality.

❑ M_w of LEL was 181K, 30 times larger than that of individual chain (~5.9K), suggesting the occurrence of aggregation.

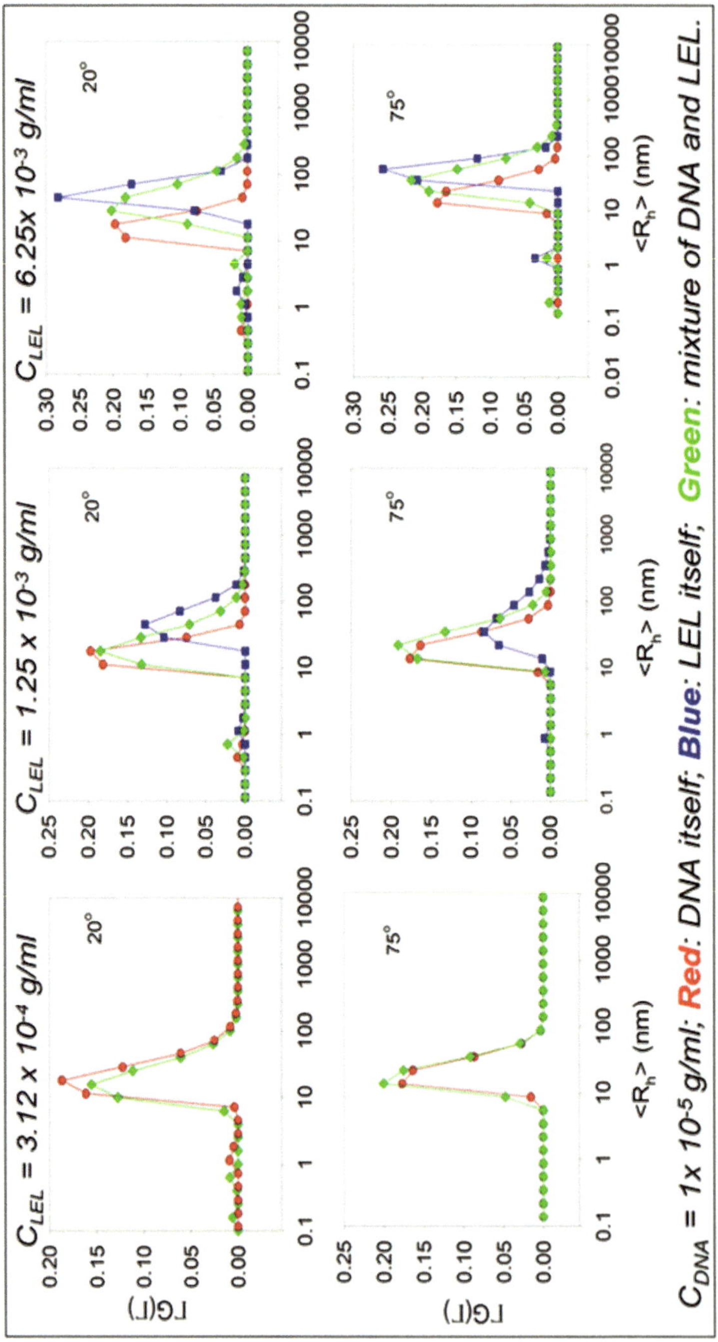

Figure 4.26 Laser light scattering on mixtures of DNA and $L_{10}E_{78}L_{10}$.

Table 4.2 Laser light scattering on mixtures of DNA and $L_{10}E_{78}L_{10}$: physical-chemical properties

C_{LEL} (g/ml)	0.0	3.12 x 10^{-4}	1.25 x 10^{-3}	6.24 x 10^{-3}
$<R_h>$ (nm)	16.9	16.5	20.6	30.7
$<R_g>$ (nm)	15.9	19.7	25.9	36.9
$<R_g>/<R_h>$	0.94	1.2	1.3	1.2

Note:

- Only one fragment was observed, indicating encapsulation of DNA.
- The size of the complex increases with increasing LEL concentration.
- Core-shell structure maintained.

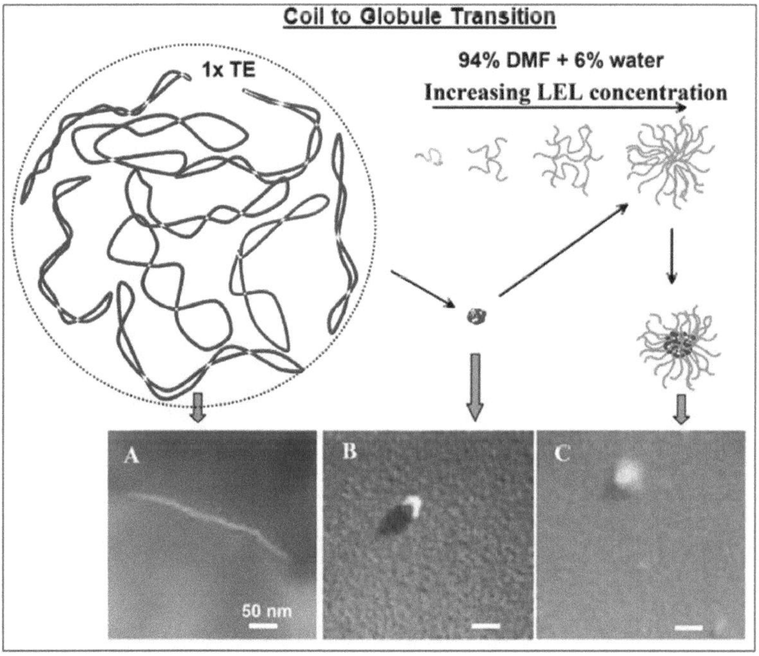

Figure 4.27 Coil-to-globule transition.

Figure 4.27 shows an illustration of DNA chain conformation. In 1× TE buffer, the DNA molecules exist as individual (somewhat extended) polymer chains while in 94% DMF + 6% water, DNA chains collapse into a globule. For $L_{10}E_{78}L_{10}$ tri-block copolymer (abbreviated as LEL), LLS results show that aggregation occurs with increasing $L_{10}E_{78}L_{10}$ concentration and eventually forming polymeric micelles which are more hydrophilic in the core. Then, the collapsed DNA chain prefers to enter into the more hydrophilic domains. To strengthen this idea, the electron micrographs under the three conditions shown in Figure 4.27 A, B, C appear to confirm the findings.

The details have been presented in [4]. We show some electro-spun scaffolds and carried out the adhesion and proliferation of MC3T3 cells within a nanostructured electro-spun PLGA scaffold. Verification of the integrity of released DNA from a 30% PLGA/BC scaffold by using agarose gel electrophoresis is shown below.

Finally, the released DNA was tested for its integrity. See Figure 4.33.

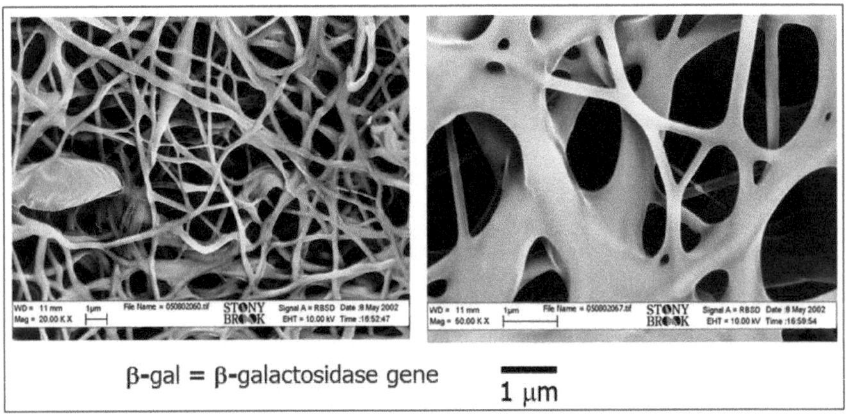

Figure 4.28 Biodegradable electro-spun scaffolds containing PLGA (90% PLGA, 10% BC)/β-gal plasmid DNA.

The scaffold (Figures 4.28, 4.29) has a suitable morphology that will enable endogenous cells to migrate into it, adhere, proliferate, and initiate new matrix/osteoid production (Figure 4.30).

Viable cells were determined by the incorporation of a dye that indicated an active metabolism 3 days after plating onto a PLGA nanostructured scaffold, as being demonstrated in Figure 4.31.

A part of Figure 4.31 is also shown in Figure 4.33 (upper left corner). Figure 4.32 shows transient *in-vitro* transfection of osteoblastic MC3T3 cells by DNA released from a biodegradable scaffold.

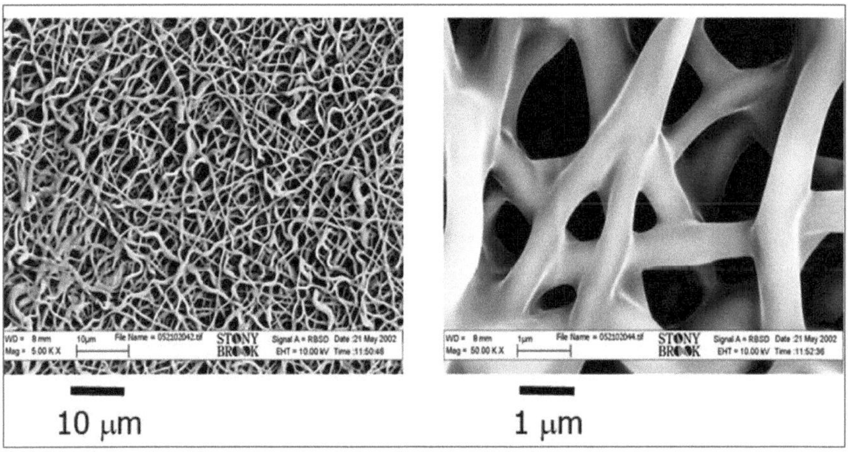

Figure 4.29 Biodegradable electro-spun scaffolds containing PLGA (85% PLGA, 15% BC)/b-gal plasmid DNA [4].

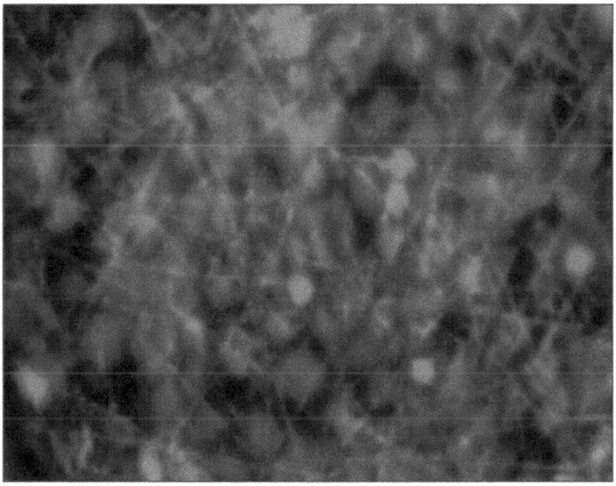

Figure 4.30 Adhesion and proliferation of MC3T3 cells within a nanostructured electro-spun PLGA scaffold.

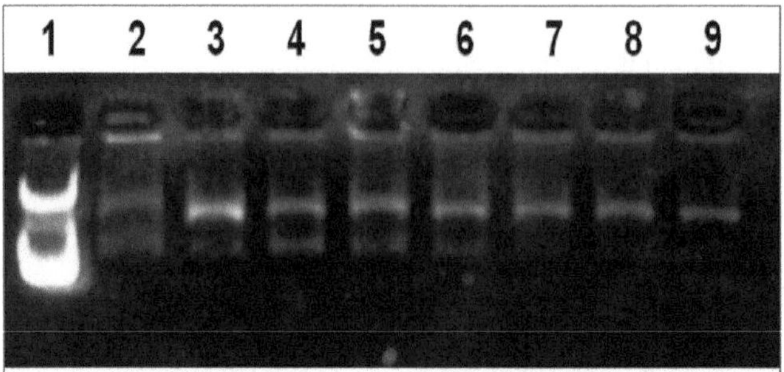

Lane 1, control (700 ng); lane 2, 2 hrs (168 ng); lane 3, 6 hrs (201 ng); lane 4,12 hrs (135 ng); lane 5, 1 d (168 ng); lane 6, 2 d (155 ng); lane 7, 3 d (111ng); lane 8, 5 d (100 ng); lane 9, 7 d (85 ng).

Figure 4.31 Adhesion and proliferation of MC3T3 cells within a nanostructured electro-spun PLGA scaffold. Verification of the integrity of released DNA from a 30% PLGA/BC scaffold by using agarose gel electrophoresis.

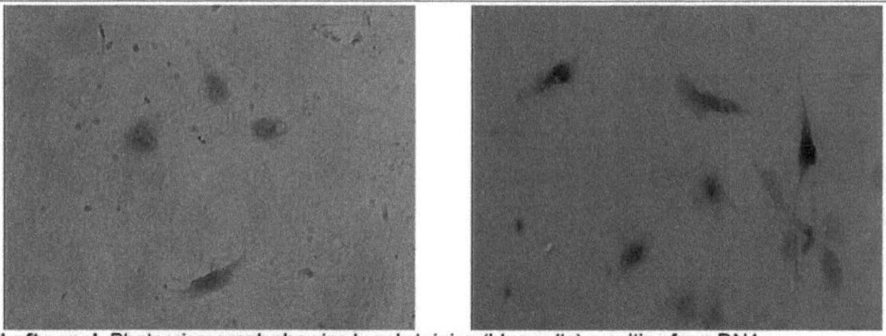

Left panel Photomicrograph showing b-gal staining (blue cells) resulting from DNA released directly from the scaffold.
Right panel Positive control test of DNA transfection using a DNA/Fugene combination

Figure 4.32 Transient *in-vitro* transfection of osteoblastic MC3T3 cells by DNA released from a biodegradable scaffold.

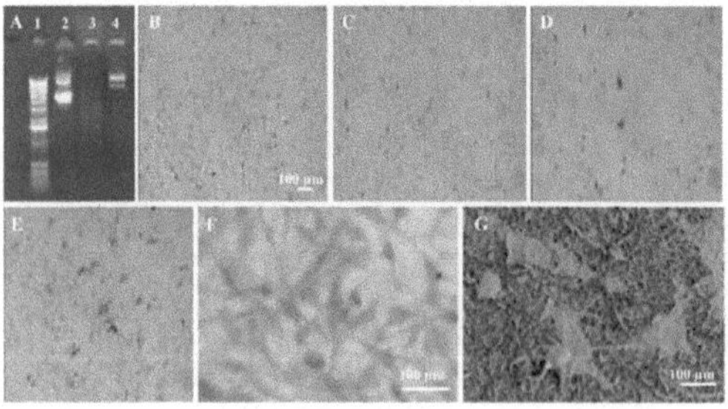

A: DNA integrity after release
1. DNA ladder
2. Plasmid DNA
3. Direct loading of DNA
4. With LEL surfactant

B: Transfection of MC3T3 cells using naked DNA directly added to cell medium

C: No transfection.

D: Transfection was observed using (A.4), i.e., with DNA encapsulated in LEL during electro-spinning and subsequent release from the scaffold.

E: Transfection with DNA using Fugene 6.

F&G: Cellular adhesion and growth on PLGA scaffold containing encapsulated DNA.

Figure 4.33 Transfection with plasmid DNA using PLGA scaffold.

Figure 4.33 summarizes the transfection of plasmid DNA by using a PLGA scaffold.

4.3.4 Summary

We summarize the findings as follows:

- In 94% DMF + 6% water, plasmid DNA molecules were condensed and aggregated.
- Contractions were of the order of $\sim 10^4$.
- LEL forms aggregates with one domain rich in PEG and water. Upon mixing with plasmid DNA, DNA, PEG and water formed a more hydrophilic domain and induced "micelle" formation.
- Capacity and size of hydrophilic domain increased with increasing LEL concentration, allowing more DNA molecules to be incorporated into the hydrophilic domain.
- TEM results confirm the above interpretations.
- DNA was protected during electro-spinning.
- Transfection was observed with DNA released from scaffold containing LEL.
- Optimization yet to be carried out.

4.3.5 Final Note

In this section, the author (BC) has used two examples to illustrate that the technologies discussed in Chapter 2 (laser light scattering, synchrotron X-ray scattering) and Chapter 3 (electro-spinning/blowing) can indeed be used to investigate a biomedical topic of interest and importance. The examples combine the knowledge from Chapters 2 and 3 that can provide not only useful applications but also permit us to achieve a more fundamental understanding of the topic under study.

Please note that toward the end, we have brought in a range of other techniques to confirm our findings and have confirmed successful findings of the study. The reader should be aware that in such studies, there will be failures in the process and we should not be afraid of failures. Instead, we should learn from failures.

4.4 Water Purification with Fibrous Membranes

4.4.1 Brief Outline

This section uses the water purification theme, with elaboration on fibrous membranes, paying attention to both synthetic and natural nanofibers. The water project is a team effort. Water is a global issue and the approach emphasizes two themes, i.e., (1) the use of fibrous format, in combination with renewable and sustainable natural materials, such as cellulose; and (2) the use of nanofiber/polymer matrix inter-surface as directed water channels.

Water challenges are based mainly on three challenges: (1) too many people, (2) global climate change, and in combination with (3) uneven water distribution as well as a lack of available fresh water. Scientifically, the fundamentals of the water molecule, H_2O, are still in progress.

By following those three challenges, the use of fibrous format can provide one approach with technology breakthroughs for a better understanding. Examples and instrumentation are presented to show the type of information available by using cellulose-type materials as a basic and significant example.

Applications to waterborne diseases caused by bacteria, viruses, and heavy metals are discussed, showing TEM images of ultrafine cellulose nanofibers, before and after adsorption of uranium, U (VI).

For conditions on the use of interfaces for directed water channels, the demonstration of the concept was carried out by using cross-flow ultrafiltration measurements of cellulose/PAN membranes for oil/water emulsion. The enhanced *directed water channel concept* are presented with experimental data.

Membrane performance can be enhanced by using slot-die coating. Furthermore, the density of the cellulose nanofibrous top layer could be tuned, for example, by cross-linking poly(ethylene glycol) diacrylate (PEGDA) within the cellulose nanofibrous network. Manipulation of top-layer properties can yield higher fouling resistance and better recovery ability.

A summary on the technology advantage is presented below.

Patents and patent applications on electro-spinning and membranes as discussed over a one-year period in 2013 are listed, with support acknowledgements from the government and industry as shown in Section 4.4.

Figure 4.34 shows an example of the range of fiber diameters used in water purification membranes.

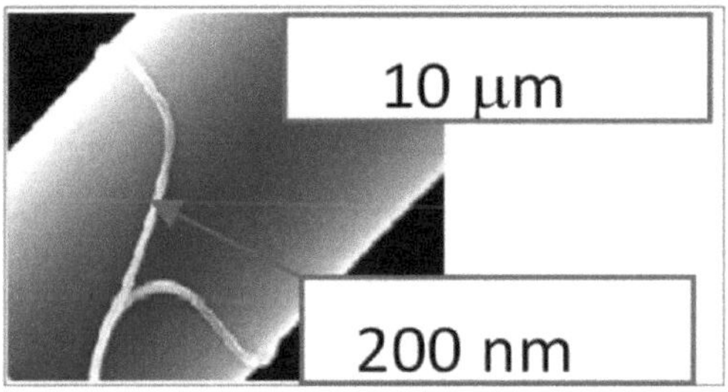

Figure 4.34 An example of the range of fiber diameters used in water purification membranes.

4.4.2 Using Water Theme with Emphasis on Global View and Water Purification

Use of Synthetic versus Natural Nanofibers

It is important to point out our preference on the use of natural plant-based nanofibers. The studies involved a research team, which, aside from Chu at Stony Brook University, consists of Benjamin S. Hsiao, Christian Burger, DF Fang, HY Ma, Ran Wang, Lixia Rong, Xiao Wang, Zhe Wang, Rui Yang, Ying Su, Anna Sato and other student participants and research staff members, over a number of years.

Motivation: Water is a global issue. We have ideas that may be worthy of further exploration:

1. Use of fibrous format for separation[a]
2. Use of nanofiber/polymer matrix interface as directed water channels
3. Take advantage of the most abundant renewable material, cellulose (crystals) as a basis component.

[a]Long Island Technology Hall of Fame awarded Patent 8,231,013 B2 "Articles Comprising a Fibrous Support" (by Chu, Hsiao, Yoon): the Long Island Patent of the Year Award in the category of Industrial Innovation on March 6, 2013 – **one of 3 from ~1500 patents.**

Challenges that have initiated the studies:

Challenge 1 (Too Many People)

If we examine the world population from 1950 to 2022, it was ~2.5 billion in 1950, reaching ~7 billion in 2011, and up to about 10-12 billion before the end of the 21st century. While the population increase in North America, Europe, and Latin America was relatively mild, Africa and Asia showed a remarkable increase, with estimates of ~9 billion by 2050.

We know that there are many people in this world, making the high density population a challenge which we need to pay attention to. Looking at challenge 1, we note that the problem becomes serious with only a 70-year time span from 1950.

Challenge 2 (Rivers Running Dry and Pollution)

Challenge 2 is on global climate change. Going back to the April 2008 issue of *Physics Today*, we already saw the receding levels in Lake Powell dropping down to less than 50% of its capacity in 1999 and Lake Mead in Nevada and Arizona was only half full. In reading the tree rings, we know that drier than the average period is coming our way. In the *National Geographic Magazine* (February, 2008), we have learned that drought in the US West, from drier than average to wetter than average is coming. The conclusion is based on data over a period from 800 to 2006 AD, a historic average over the past 1200 years. It is indeed an amazing feat achieved by patient and dedicated scientists reading the tree rings.

Currently, we know that many parts of the United States are feeling the water shortage challenge.

In recent years, Western Europe has done more in reducing its greenhouse gas emissions than other parts of the world, including the United States and Japan as well as countries with large land areas, such as Australia and Canada. Yet, for example in 2022 summer, Europe is experiencing an extremely hot summer, e.g., in London and Paris, while the United States has excessively high temperatures in many parts of the country and historical rainfalls in the Mid-West. Thus, climate change has no boundaries. It is a world issue, with burdens falling heavier on poor people.

Challenge 3 (Limited Available Fresh Water)

Challenge 3 for the water problem is the global water distribution and the availability of fresh water in this world. In the distribution of water, ~96.5% of total water is in the oceans. Out of ~2.5% fresh water, about 69% is in Greenland and Antarctica and a large amount of the remaining fresh water is under the ground, leaving surface water, such as rivers, and other fresh water sources to the amount of ~20% in lakes and over 70% in ice and snow. To summarize, rivers contain 1–2 parts of fresh water per million of total water in this world [5] and the world population will be about 9 billion by 2050.

Beyond the infrastructure, projects, such as building dams and diverting water courses in order to enhance water security, are very challenging and costly undertakings. However, some projects may also try to do too much to protect the biodiversity of associated ecosystems which we should not ignore. At the end, water security increases with affluence but may also affect biodiversity.

The magnitude of some of those projects in China is huge, as China needs water, especially for parts of northern China. The South-North water diversion project has many challenges. One observation is that the Chinese people already use ~25% of water per person when compared with Americans. Yet, China has about 4 times the population and only about half the amount of water when compared with that in the United States. Thus, the water challenge is also serious for China.

There is no simple solution to the water purification challenge. Furthermore, conditions in different locations are different. Nevertheless, water purification is a very important process which our society needs. It should also be recognized from the very beginning that the use of nanofibers, being fabricated by means of the electro-spinning/-blowing processes for creating a useful and critical component in the filtration membrane, is only *one* of the useful pathways, worthy of our attention and consideration.

In water purification, we often think of everything associated with water; but have taken for granted what the water molecule (H_2O) is. In reality, we do not know as much as we think we do, especially when we come close to the molecular level and start to think of its interactions with molecular surfaces and in cages, such as a 60-carbon-atom cage. Thus, we may also wish to broaden our thinking pathway as the fundamentals on water are still in progress.

Let us now provide a brief discussion on the current **technology breakthroughs.**

Figure 4.35 illustrates the three-layer concept of sizes of fibers and the three electron microscopy images on the right show how the three layers look. A very thin top layer partially interpenetrated by thin fibers of the second layer is responsible for the separation of molecules or particles depending on the type of membranes we want to produce, such as membranes for microfiltration (MF), ultrafiltration (UF), nanofiltration (NF), and reverse osmosis (RO). It turns out that the use of fibrous structures has advantages in increasing the throughput, as we have already discussed in Chapter 3.

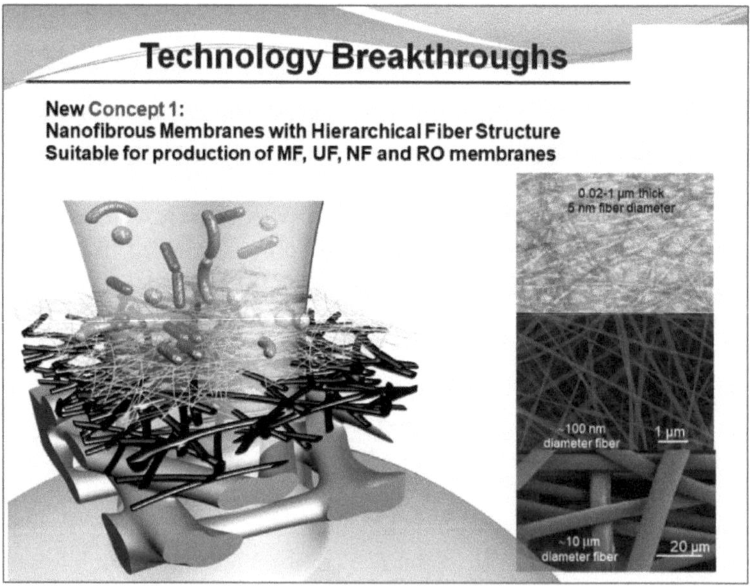

Figure 4.35 Technology breakthroughs.

The advantages are further illustrated in Figure 4.36. The separation capabilities of the conventional membrane and the nanofibrous membrane of similar size are compared by using electron micrographs for both sectional and surface views. The filtration efficiency for the nanofibrous membrane was 3 to 10 times better.

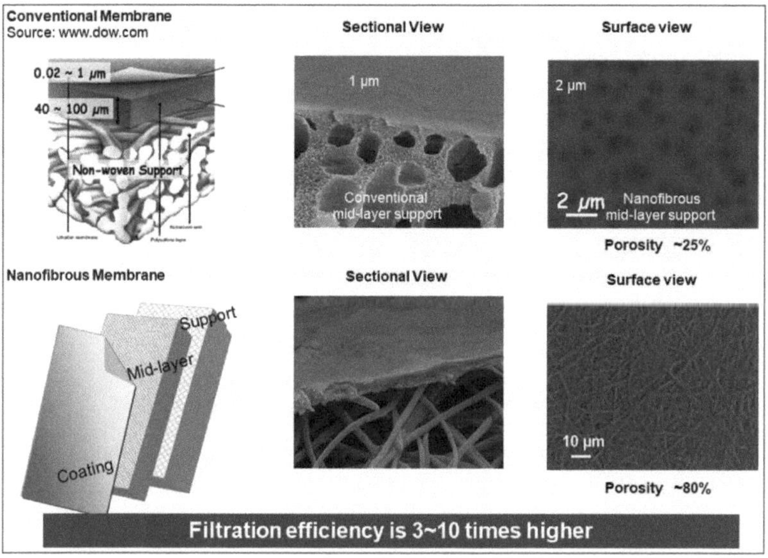

Figure 4.36 Advantage: Superior Porosity—better permeability.

The advantages of the fibrous format for separation membranes can further be illustrated by considering inter-connected continuous free space and soldering-like attachments (Figure 4.37). In comparison with woven fabrics (Figure 4.37a), the holes of non-woven fabrics (Figure 4.37b) over multiple layers show a flow-through format that is somewhat different from those of a woven fabric. The volume of empty space for non-woven fabrics can easily be of the order of 80%. To preserve the integrity of the non-woven fabrics, we can "solder" (or anneal, meaning by means of chemical or/and physical connections) the junctions together, as shown schematically in Figure 4.37c. Thus, such nanofibrous scaffolds can be fairly strong, with *higher flux and lower fouling properties at the same time.*

Figures 4.37 and 4.38 came from Schematic Drawings of Woven & Non-Woven Fabrics, as done by Professor Christian Burger in our research team at Stony Brook. The approach of using non-woven fiber staking can further be shown by examining the fiber diameter and pore size in non-woven membranes (Figure 4.38), where we see the fiber diameter ratio varying in 1:3:10 with the same porosity of 80%.

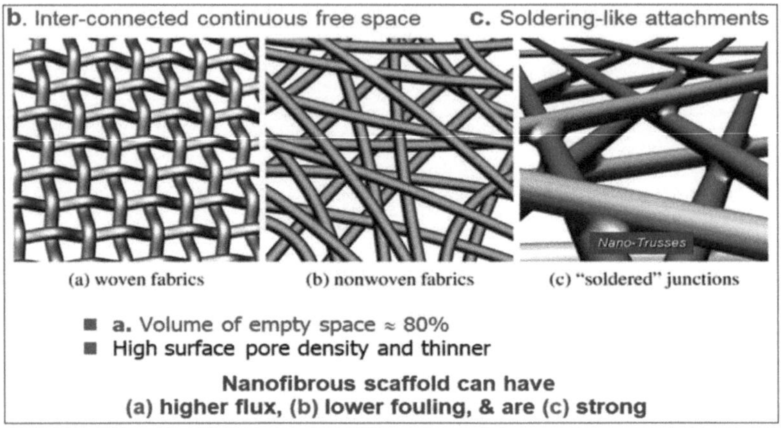

Figure 4.37 Advantages of fibrous format for separation membranes.

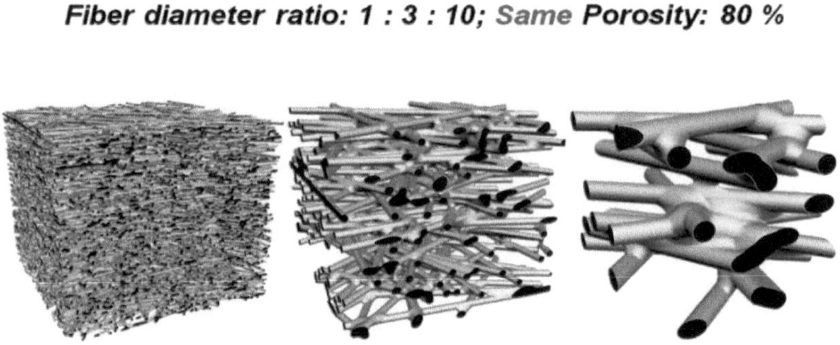

Figure 4.38 Fiber diameter and pore size in non-woven membranes [6].

The effects have been further demonstrated by providing the following set of experiments:

1. Fiber diameter has little effect on porosity (Figure 4.39).
2. Low hydraulic resistance of poly-acrylonitrile (PAN) E-spun membranes (Figure 4.40).
3. Fiber diameter and pore size relationships have been determined (Figure 4.41) [6].

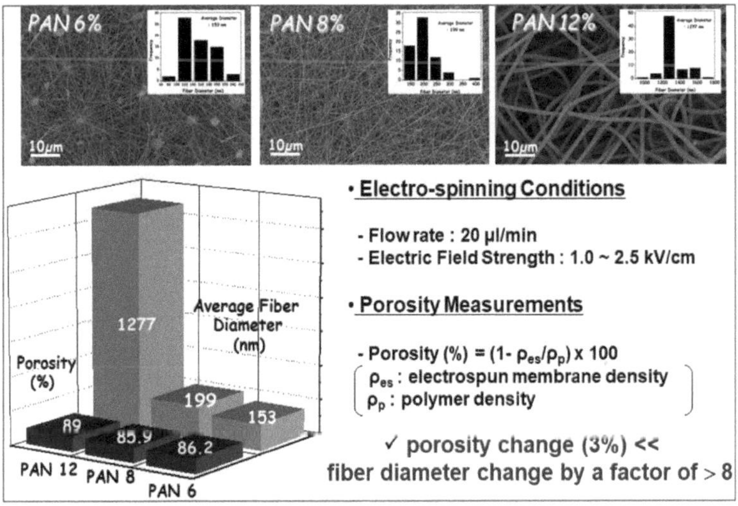

Figure 4.39 Fiber diameter has little effect on porosity.

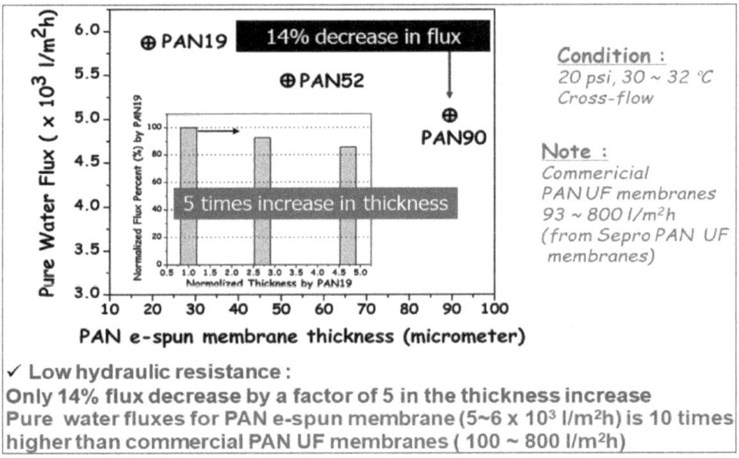

Figure 4.40 Low hydraulic resistance of poly-acrylonitrile (PAN) E-spun membranes.

The more efficient water purification process has provided us with a new scheme on water purification for waterborne diseases (Figure 4.42). Based on bacteria sizes of most common waterborne diseases, such as dysentery, cholera, and typhoid (information due to Dr. Laurence Chu, MD), we can design efficient filtration membranes with a pore size of ~300 nm and a pore volume of ~85%, meaning a high throughput membrane for taking out those bacteria by filtration without special instrumentation. Furthermore, as has been shown, the membranes are mechanically strong with nano-trusses.

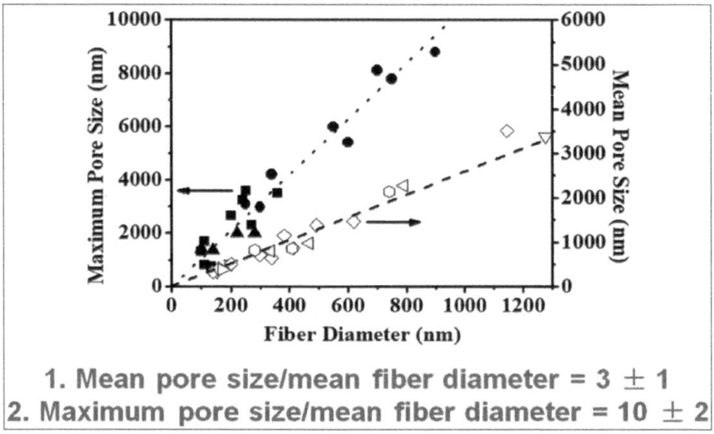

Figure 4.41 Fiber diameter and pore size relationships [6].

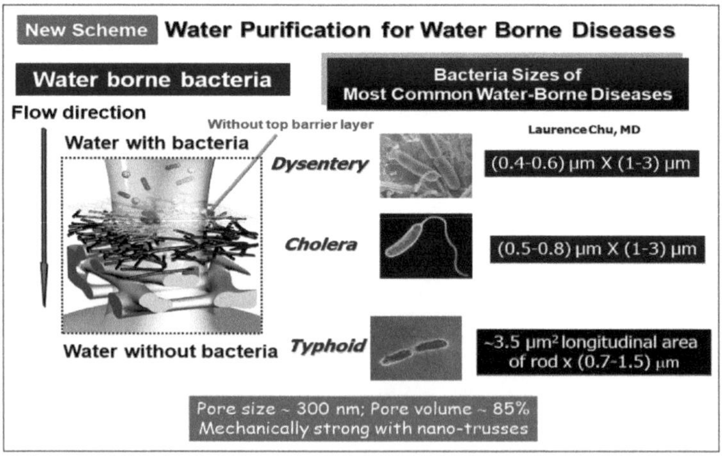

Figure 4.42 Water purification for waterborne diseases.

We can use microfiltration (MF) membranes for the removal of bacteria and viruses. As shown in the figure, and for typical pore sizes of MF membranes, there are approaches for separations based on size exclusion and adsorption with the typical contaminants also indicated in the figure on Microfiltration

Membranes for Removal of Bacteria & Viruses. In the Figure, the typical pore size ranges for microfiltration (MF) membranes are sufficiently large for easy fluid passage without much applied pressure and yet sufficiently small to exclude unwanted bacteria, such as those for dysentery, cholera and typhoid, being related to waterborne diseases.

There are many alternatives in the choice of filtration membranes. Aside from pore size, there are, for example, surface adsorption and charge effects. The key considerations often deal with the specifications on the impurities in the fluid system and the environment associated with the type and amount of impurities that need to be removed.

Useful Application 1:

The microfiltration performance of a few E-spun filters was compared with Millipore GS and RA filters in terms of average flux, maximum pore size, and average pore size by using a dead-end flow system at 2.28 psi by gravity (Figure 4.43). Note the average flux and maximum pore size for the E-spun PVA/non-woven PET filter in comparison with the Millipore GS filter. The increase in the average flux shows that the E-spun approach can produce an efficient filter for the removal of bacteria and viruses, such as *E. coli*, hepatitis A virus, and heavy metal ions (Cr, Pb, etc.), *by gravity, without additional applied pressure.*

Microfiltration Performance

Name	Total thickness (μm)	Ave. flux (1000L/m^{2}h)	Max. pore size (μm)	Ave. pore size (μm)
E-spun PES/Coffee Filter Paper	130	16.6	2.5	0.8
	140	13.4	1.7	0.6
	155	11.7	1.2	0.4
E-spun PVA/nonwoven PET	**160**	5.5	**0.6**	**0.2**
Millipore GS 0.22 μm	**175**	0.39	**0.6**	**0.2**
Millipore RA 1.20 μm	145	2.0	4.4	1.5

* All the above test were processed on the dead-end flow system at 2.28 psi by gravity.

Figure 4.43 Microfiltration performance.

Useful Application 2:

The second application is based on the advantage of our **cost-effectiveness green approach.** It deals with the use of cellulose nanofibers. Cellulose nanofibers can be prepared with diameters down to ~5 nm, *without* using the electro-spinning process. The cellulose-type fibers not only reduce the cost in producing nanofibrous membranes but draw the source more directly from nature.

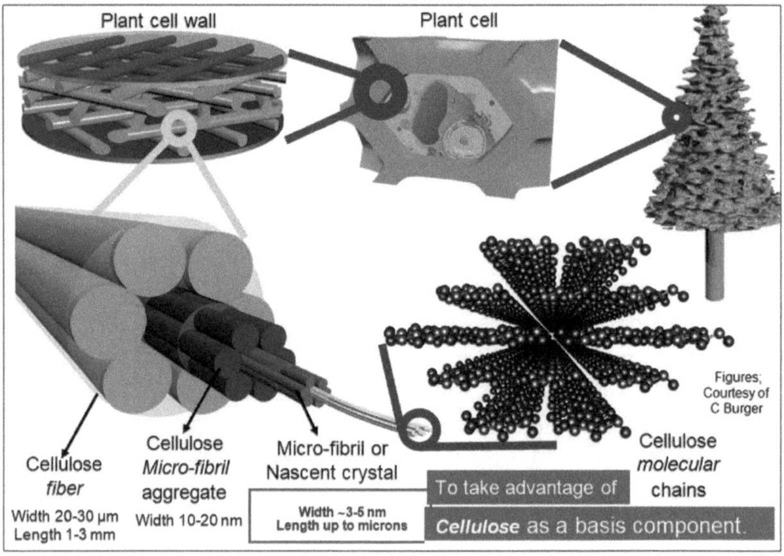

Figure 4.44 Hierarchical structure of wood biomass.

Figure 4.44 illustrates schematically how we can take advantage of cellulose as a basis component, starting from the tree and going down to cellulose molecular chains with widths of a few nanometers [7].

The preparation of ultrafine cellulose nanofibers with fiber diameter of ~5 nm is sketched in Figure 4.45.

An outline of the experimental process, including the type and source of raw materials as well as an outline of the process is presented in Figure 4.46.

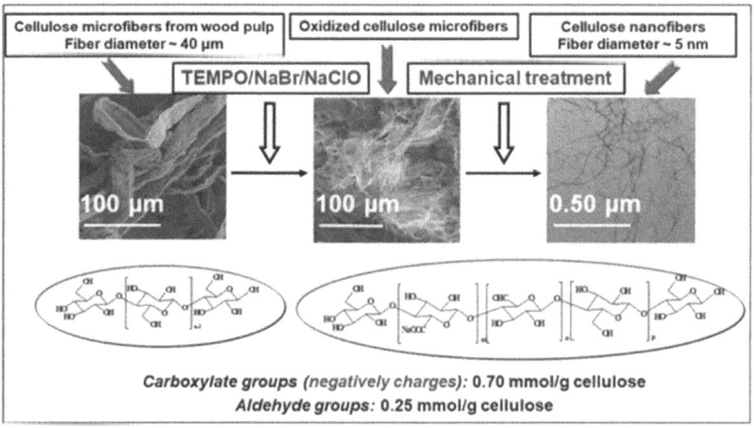

Figure 4.45 Preparation of ultrafine cellulose nanofibers.

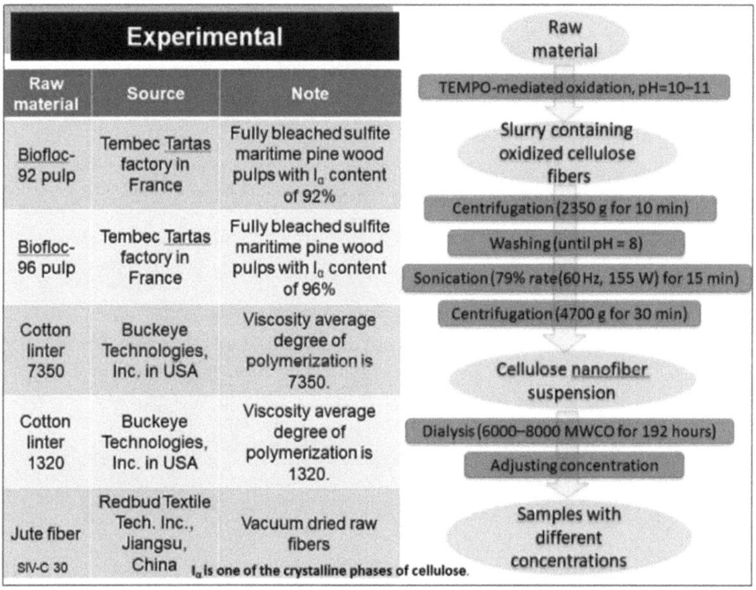

Figure 4.46 Preparation process of ultrafine cellulose nanofibers.

The detailed analysis by means of SAXS/WAXD and model computations, including the use of different cylindrical rod or ribbon models and the Debye Function analysis, can yield useful information on cellulose nanofibers from different sources, such as cotton and jute, etc. For the current discussions, with the physics basis being of primary interest, we have used electron microscopy as the main tool to support our findings on the performance as related to the water purification process. More experiments on microfiltration membranes impregnated with cellulose nanofibers and cellulose nanofiber-based membranes for oil/water emulsion, together with electron micrographs, are presented in Section 4.5.

TEM images of polysaccharide nanofibers can clearly show the magnitude of the diameters of those nanofibers of interest (Figure 4.47). There is a range on the size estimate of those natural nanofibers. Such information is sufficient as a guideline to fabricate separation membranes for specific applications.

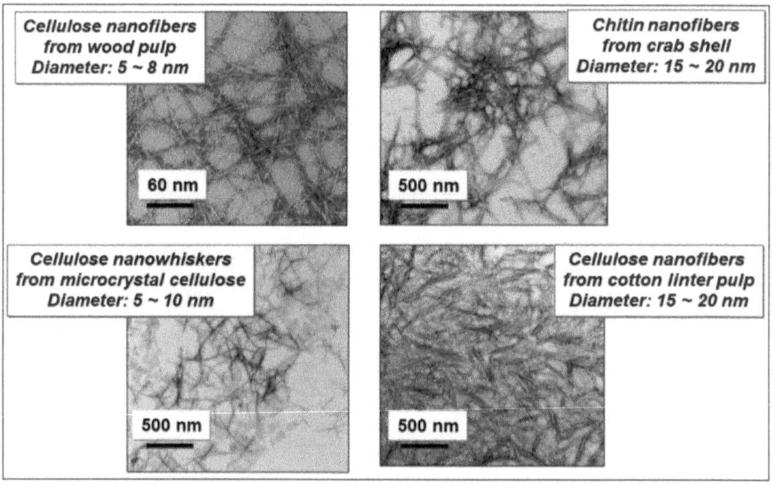

Figure 4.47 TEM images of polysaccharide nanofibers [8].

Now, we expand our range of waterborne diseases caused by bacteria, viruses, and heavy metals.

Examples for bacteria are *E. coli* and Leptospirosis, for viruses are SARS and hepatitis, and for heavy metals are arsenic (As III & V) and chromium (Cr (VI)) with the primary methods of separation being shown as size exclusion for bacteria, adsorption by charge interactions for viruses, and adsorption by charge interactions and chelating agents for heavy metals, as shown in Figure 4.48.

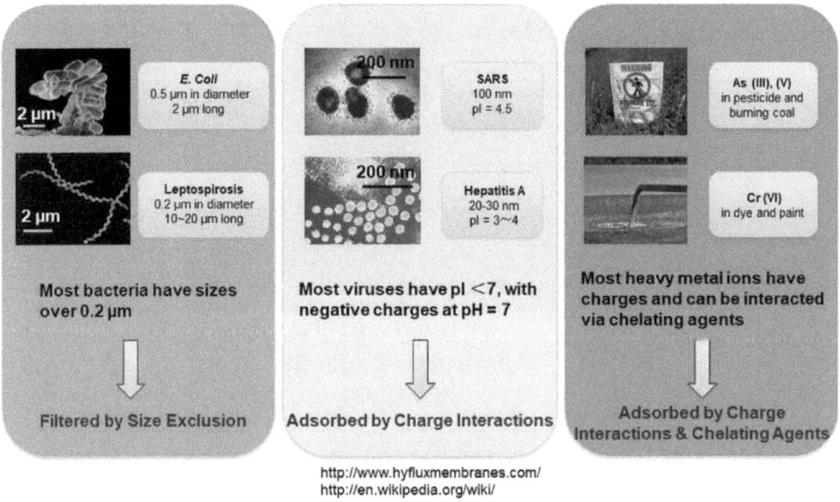

Figure 4.48 Waterborne diseases caused by bacteria, viruses, and heavy metals.

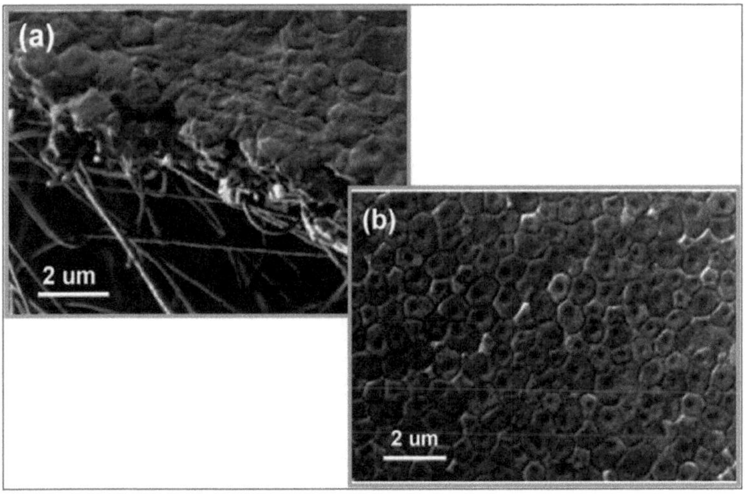

Figure 4.49 Cellulose nanofibers MF membrane for removal of *E. coli* by size exclusion. *E. coli* was covered on the surface of microfiltration membrane, and the retention was 99.99% [9].

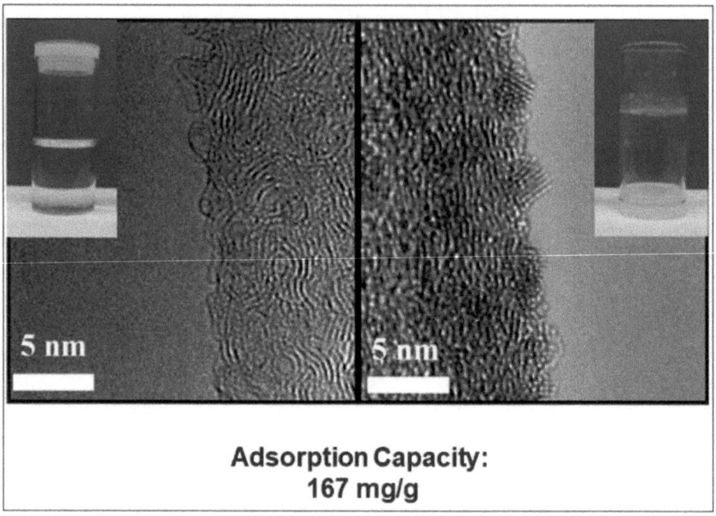

Figure 4.50 TEM images of ultrafine cellulose nanofibers before and after adsorption of U(VI).

For more details, see [10].

Figure 4.51 shows cross-flow ultrafiltration measurements of cellulose/PAN membrane for oil/water emulsion. The permeation flux of cellulose/PAN membrane was about 10 times higher when compared with that of PAN10, with essentially the same rejection ratio (hollow circles and hollow squares).

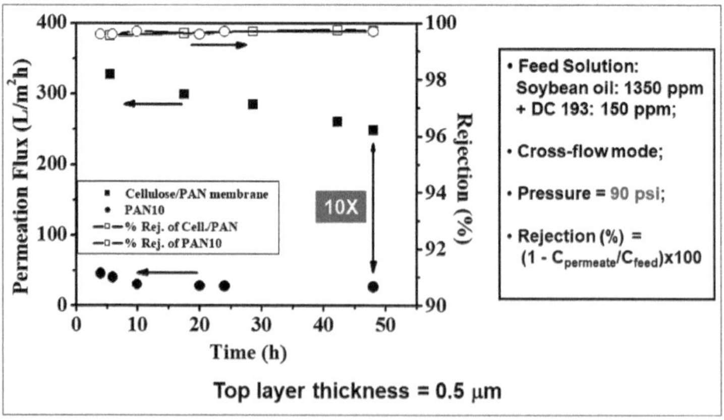

Figure 4.51 Water channels. Cross-flow ultrafiltration measurements of cellulose/PAN membrane for oil/water emulsion.

The experiments simply try to demonstrate the efficiency in our deployment of nanocellulose fibers. We try to provide our thoughts on how **functional nanofibers can enhance the performance of directed water channels**, as shown in Figure 4.52 and depicted by the corresponding electron micrographs for the base support, the nanofibrous mid-layer, and the thin top layer. The fiber surface with hydrophobic and hydrophilic regions forming the surface layer (in (blue) dark color) acting as water channels.

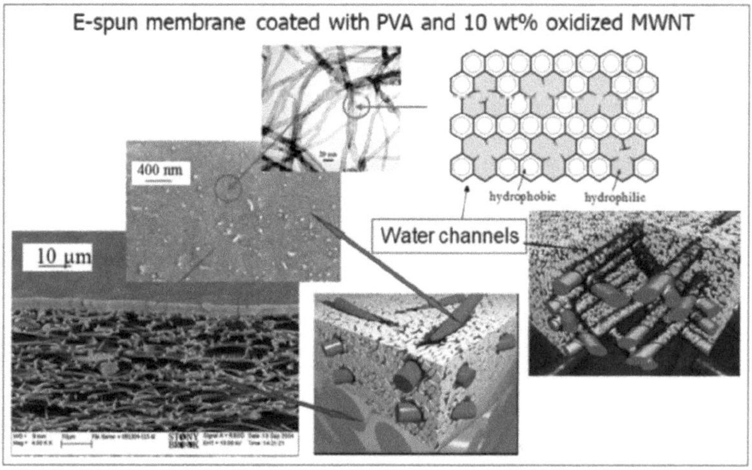

Figure 4.52 Functional nanofibers to enhance directed water channel.

Emulsified oily water filtration was used as a test for demonstration of the concept (Figure 4.53). Note the coating layer thickness difference (3 μm versus 100 μm) and the difference in flux of the two systems.

The membrane efficiency as enhanced by the use of directed water channels has been demonstrated experimentally (Figure 4.54). We now attempt to provide a physical picture, with support from SEM images of the NF membrane with CN as substrate. The figure (4.54) on upper left shows a side view of the SEM image, including the top (CN) layer and the electro-spun mid-layer. The two (A and B) panels on the right show the expanded views of the top (A) layer and the lower (B) electro-spun layer without the cellulose nanofibers.

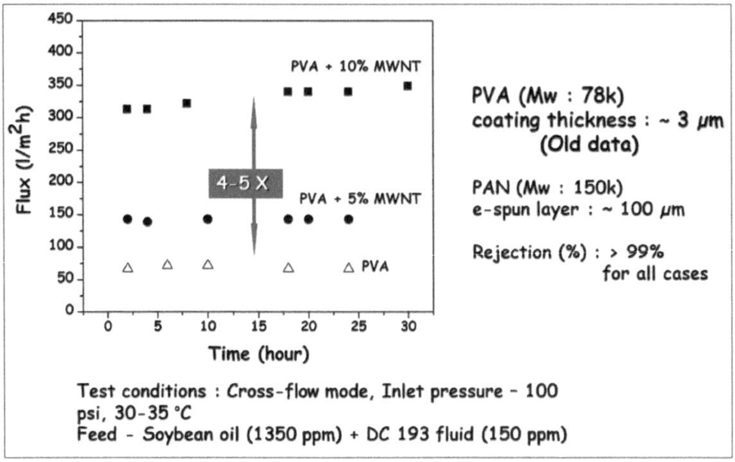

Figure 4.53 Emulsified oily water filtration: PVA/MWNT-PAN for 24 hours.

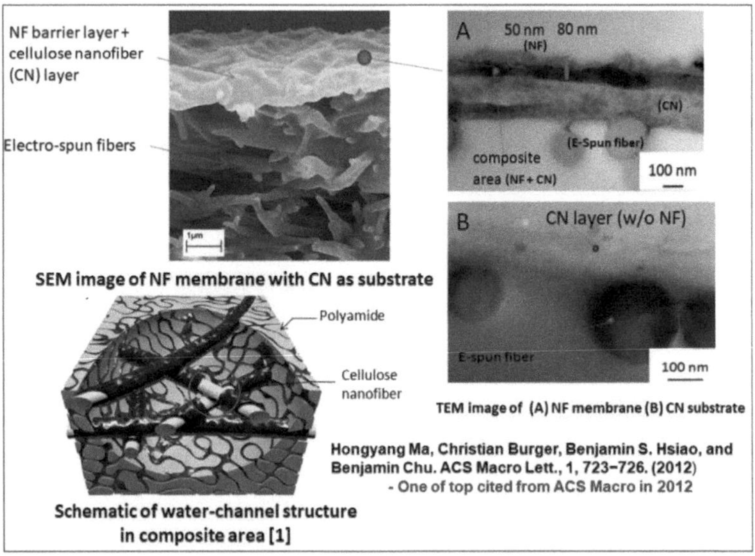

Figure 4.54 Membrane efficiency as enhanced by the use of directed water channels [11].

A schematic diagram of the water-channel structure in the composite area, representing the NF barrier layer with the cellulose nanofibers, as shown in the A panel, depicts how the cellulose nanofiber surface can act as water channels to enhance the flow of water through the membrane.

The membrane performance can further be enhanced by slot-dic coating (Figure 4.55). Furthermore, there are additional ways to enhance the performance, such as the cross-linking of poly(ethylene glycol) diacrylate (PEGDA) within the cellulose nanofibrous network (Figure 4.56).

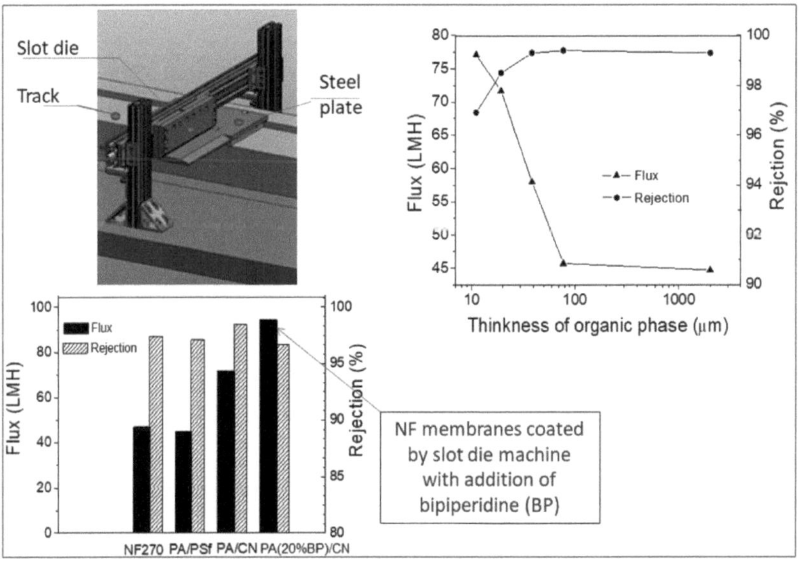

Figure 4.55 Membrane performance further enhanced by slot-die coating.

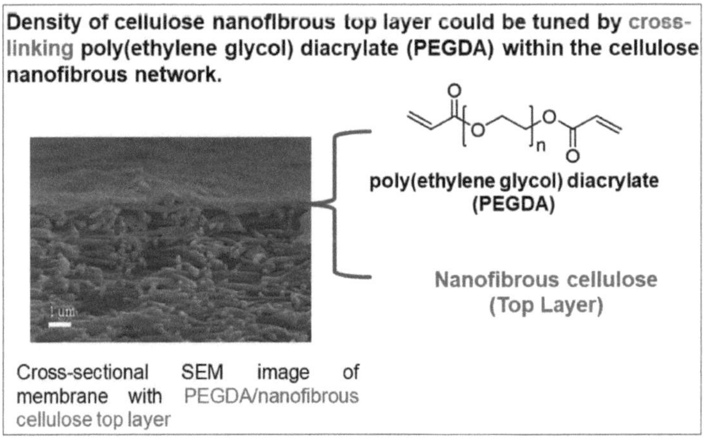

Figure 4.56 Nanofibrous cellulose composite membrane with a poly(ethylene glycol) diacrylate top layer.

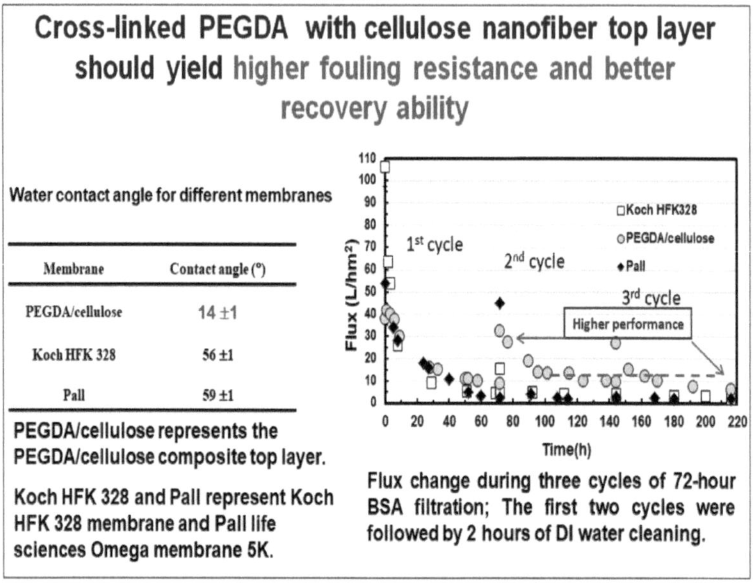

Figure 4.57 Nanofibrous cellulose composite membrane with a poly(ethylene glycol) diacrylate top layer.

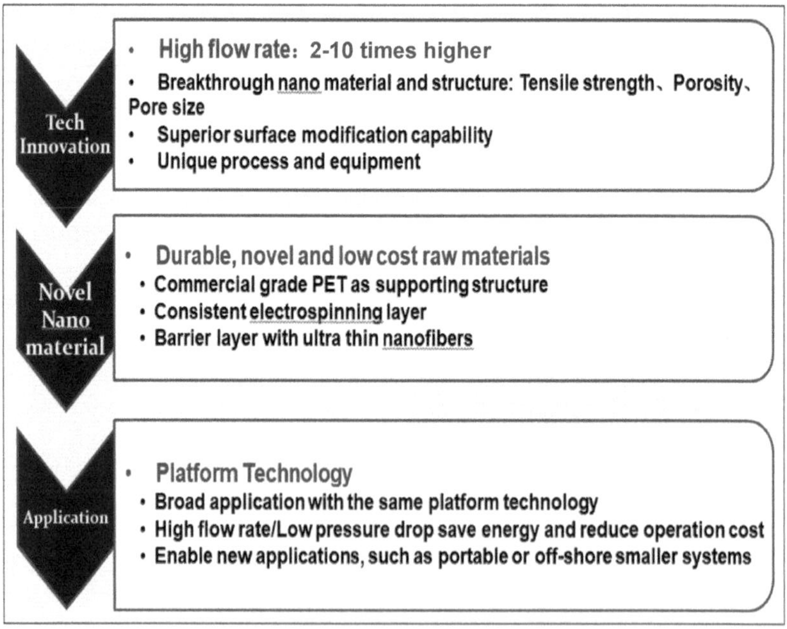

Figure 4.58 Technology: advantage; summary.

Cross-linked poly(ethylene glycol) diacrylate (PEGDA) with the cellulose nanofiber top layer yields a higher fouling resistance and better recovery ability. The test was carried out during three cycles of 72 hours each. The results showed that PEGDA with nanofibrous cellulose yielded higher performance.

E-Spinning & Membranes: Patents and Patent Applications - Chu (In addition to the patents listed in Chapter 3)

2008.

9. Benjamin Chu, Benjamin S. Hsiao, Michael Hadjiargyrou, Dufei Fang, Steven Zong and Kwangsok Kim, "Cell Delivery System Comprising a Fibrous Matrix and Cells," issued January 29, 2008, Patent #7,323,190.
10. Benjamin Chu, Benjamin S. Hsiao, Dufei Fang and Akio Okamoto, "Crosslinking of Hyaluronan Solutions and Nanofibrous Membranes Made Therefrom," issued January 29, 2008, Patent# 7,323,425.

2010.

11. Benjamin Chu, Benjamin S. Hsiao, Dufei Fang and Akio Okamoto, "Electro-Blowing Technology for Fabrication of Fibrous Articles and Its Applications of Hyaluronan," issued February 16, 2010, Patent #7,662,332.

2011.

12. Benjamin Chu, Benjamin S. Hsiao and Dufei Fang, "Apparatus for Electro-Blowing or Blowing-Assisted Electro-Spinning Technology and Process for Post Treatment of Electro-spun or Electro-blown Membranes," US Patent 7,887,311 (2011).
13. Benjamin Chu, Benjamin S. Hsiao and Dufei Fang, "Apparatus for Electro-Blowing or Blowing-Assisted Electro-Spinning Technology and Process for Post Treatment of Electro-spun or Electro-blown Membranes," U.S. Patent 7,934,917 (2011).

2012.

14. Benjamin Chu, Benjamin S. Hsiao, Dufei Fang and Kwang-sok Kim, "High Flux and Low Fouling Filtration Media," U.S. Provisional Patent Application No. 60/616,592, filed October 6, 2004; PCT Int. Appl. WO 2007001405 (2007), AU 2005333585 (issued by Australia), IN 240572 (issued by India, 2011), U.S. Patent 8,222,166 (2012).

2013.

15. Benjamin Chu, Benjamin S. Hsiao and Kyunghwan Yoon "Articles Comprising a Fibrous Support," filed in SUNY-Stony Brook (R-7925), Docket No. 788-77, U.S. Provisional Application Serial Nos.: 60/872,891 (August 4, 2006) and 60/873,086 (December 06, 2006); or "Articles Comprising a Fibrous Support" U.S. Patent 8,231,013 (2013). *Long Island Hall of Fame: Patent of the Year Award in the category of Innovation Industry.*

4.5 Highly Permeable Nanofibrous Membranes for Water Purification

4.5.1 High-Flux Nanofibrous Micro-/Ultrafiltration (MF-UF) Membranes

For high-flux nanofibrous microfiltration-ultrafiltration (MF-UF) membranes (Figure 4.59) and with the three-tier (thin-layer coating–mid-layer–support) approach, there are two breakthroughs: (1) increase in filtration speed and (2) reduced fouling. Moreover, ultrafine cellulose nanofibers (UfCN) (Figure 4.60) were used either (a) to infuse into the nanofibrous MF membrane or (b) coated on to the top layer of the nanofibrous UF membrane.

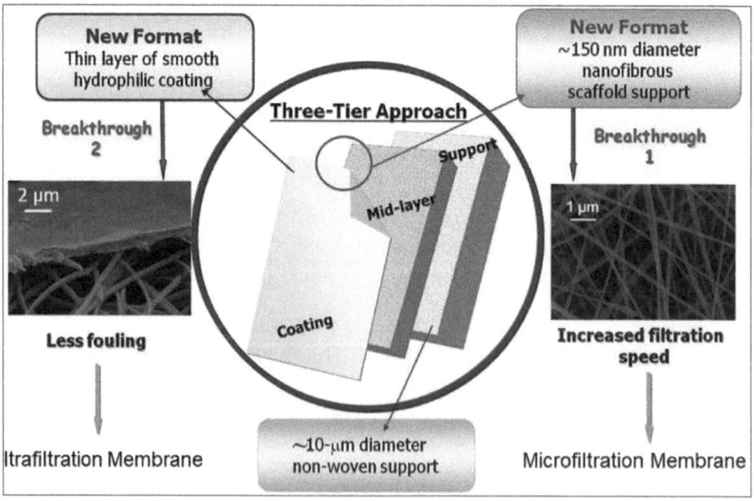

Figure 4.59 High-flux nanofibrous (MF-UF) membranes [12].

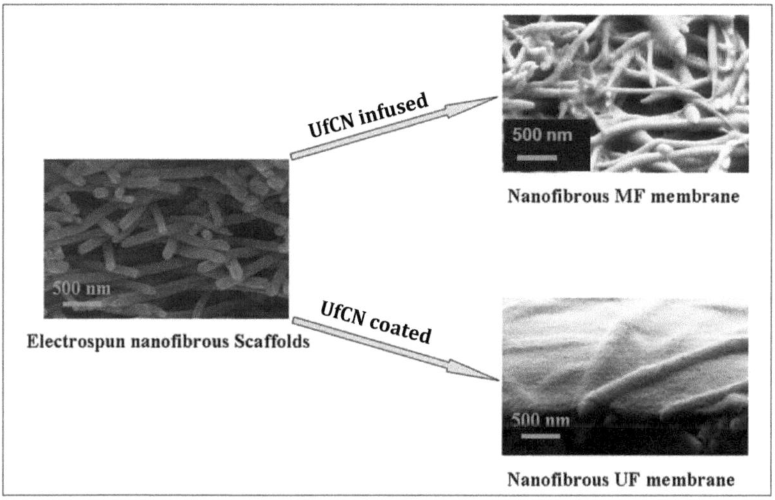

Figure 4.60 Ultrafine cellulose nanofiber (UfCN) impregnated micro-/ultra-filtration membranes. Note: please change UCN to UfCN in Figure 4.61.

The electron micrographs in Figure 4.60 show how the nanofibrous MF membrane and the UF membrane surface have been modified by the introduction of the UfCN.

The microfiltration membrane impregnated with ultrafine cellulose nanofibers (UfCN) shows the size difference between the E-spun nanofibers, as shown in Figure 4.61 a,c, and the even finer cellulose nanofibers (CNF), as show in Figure 4.61 b,d. More importantly, the finer cellulose nanofibers cover not only the pores, as shown in Figure 4.37, but also the E-spun nanofibers (with much larger fiber diameters) which act as supports for the UfCN.

Figure 4.62 shows the three-layer structure in the middle figure with the top layer containing the cellulose nanofibers (CNFs), the middle layer of E-spun fibers and the bottom support layer with a scale of 100 μm. SEM images of UfCN-based UF membrane with a barrier layer thickness of ~100 nm are shown on the right.

The inset in Figure 4.62 (top left) shows an amplified view of the upper layer with the electro-spun fibers (dark color) below the top layer and a scale of 500 nm. The electron micrographs are presented on the right side showing the side views of the upper two layers of the UF membrane, with the cellulose nanofibers being invisible (unresolved in the electron micrograph) even on the enlarged lower side view of the upper two layers of the ultrafiltration (UF) membrane. Note that the barrier layer thickness was of the order of ~100 nm.

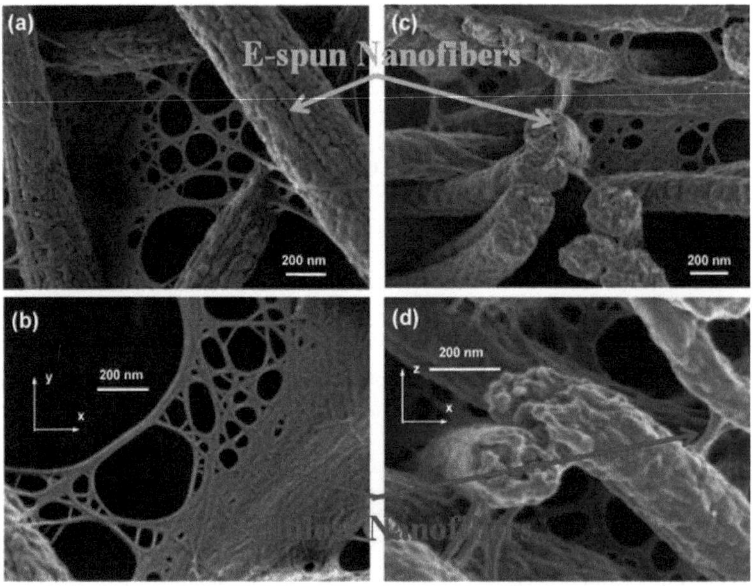

Figure 4.61 Microfiltration membrane impregnated with cellulose nanofibers [13, 14].

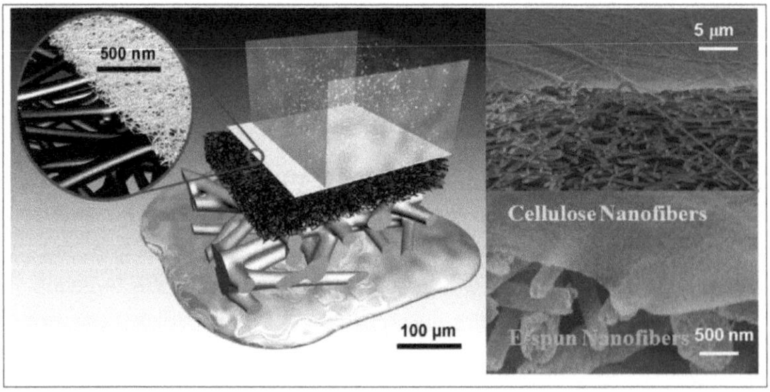

Figure 4.62 Cellulose nanofiber coated ultrafiltration membrane.

We also carried out a cross-flow ultrafiltration study of cellulose nanofiber-based membrane for oil-water emulsion. The permeation flux of the *ultrafiltration cellulose nanofiber* (UCN) based membrane was ~11× higher than that of PAN10 with a comparable rejection ratio, and ~2× higher than that of PAN400 with a lower rejection ratio.

More details are available in ref. [15].

4.5.2 UF Cartridges and Filtration Performance

Cellulose nanofiber-based UF cartridges were constructed, as shown in Figure 4.63, and the filtration performance in oily wastewater treatment was tested in comparison with ceramic UF cartridges. The permeation flux of UCN UF cartridge was ~3× higher than that of a ceramic cartridge with a comparable rejection ratio. The UCN UF cartridge could be recovered (~100%) by washing with soapy water after use. The membrane dimensions were 1.0″ OD × 15′ long with a membrane area of 1.0 ft^2.

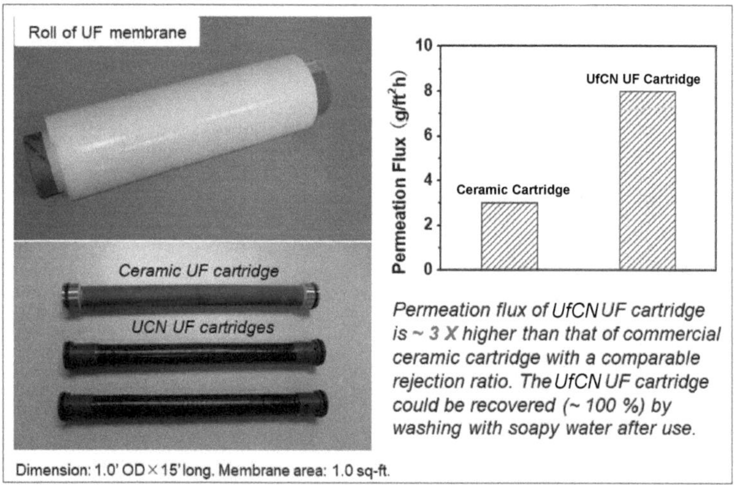

Figure 4.63 Cellulose nanofiber-based UF cartridges and filtration performance in oily wastewater treatment. Note: please change UCN to UfCN (3 places) in Figure 4.64.

4.5.3 Membrane Efficiency Enhanced by Slot-Die Coating

There are other methods to enhance the membrane efficiency, such as by using slot-die coating (Figure 4.64) or an appropriate coating method (Figure 4.65). In the coating methods, we demonstrate the results by trying out different types of coatings, including NF270 PA/CN manual coating, PA/CN slot-die coating, PA/PSf manual coating, and PA/PSf slot-die coating. The results, as shown in Section 4.5.5, revealed that the membrane efficiency was improved mainly because of controlled barrier-layer thickness.

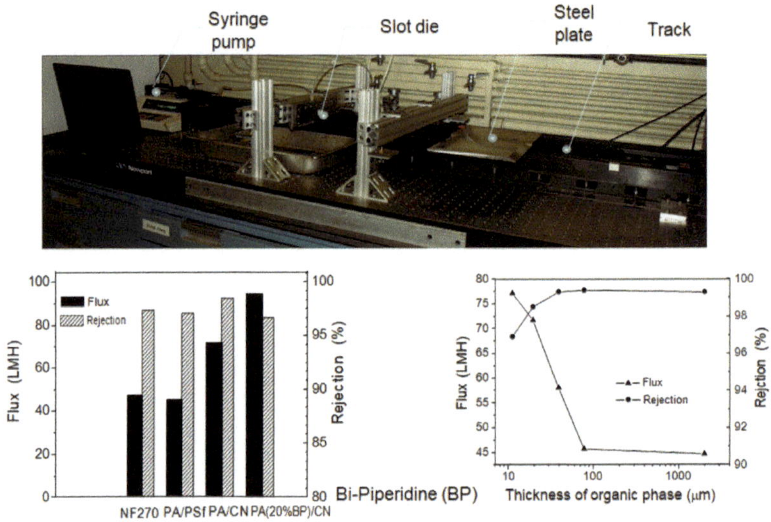

Figure 4.64 Membrane efficiency enhanced by slot-die coating.

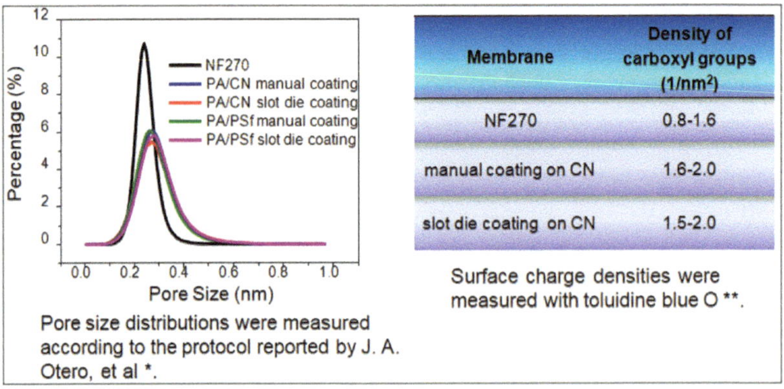

Figure 4.65 Membrane properties as influenced by coating methods. Neither pore size nor surface charge density was changed by slot-die coating. The membrane efficiency was improved mainly because of controlled barrier-layer thickness. * [16]; ** [17].

Aside from the methods presented above, there shall be new developments for further improvements. Thus, the current range of approaches merely suggests some aspects of what has been accomplished so far. There will be future improvements being developed in a variety of ways.

4.5.4 Membrane Properties as Influenced by Coating Methods

A range of membranes from different companies, such as GE, Trisep, and Koch, were compared by using a plot of rejection (5) versus permeability (L/m^2 h psi) in comparison with A1, A2, B1, B2, as shown in Figure 4.66. Permeability of our lab-made NF membrane B2 is much higher than that of commercial membranes with comparable salt rejection.

Point	Company	Membrane	Point	Company	Membrane	Point	Company	Membrane
G1	GE	DK	D1	Dow	NF90	Se1	Sepro	NF-1
G2	GE	AK	D2	Dow	NF200	Se2	Sepro	NF-2
G3	GE	DL	D3	Dow	NF270	Se3	Sepro	NF-3
G4	GE	HL	D4	Dow	NF70	Se4	Sepro	NF-4
G5	GE		D5	Dow	NF400			
G6	GE	Duraslick NFHS	D6	Dow	NF			
Tr1	Trisep	TS80	D7	Dow	XLE-400			
Tr2	Trisep	TS83	D8	Dow	LE-400			
Tr3	Trisep	XN45	D9	Dow	BW-400			
K1	Koch	TFC-S	Sa1	Saehan	RE-BLR	A1	lab-made	manual coating on Electro-spun PAN
K2	Koch	TFC-SR2	Sa2	Saehan	NE-90	A2	lab-made	manual coating on Electro-spun PAN, BP added
K3	Koch	TFC-SR3	To	Toray	UTC-70	B1	lab-made	slot die coating on CN
			H1	Hydranautics	ESPA	B2	lab-made	slot die coating on CN, BP added
			H2	Hydranautics	ESNA			

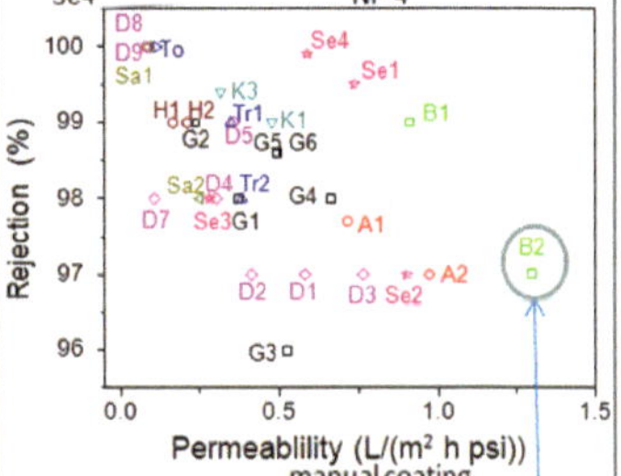

Figure 4.66 Permeability of our lab-made NF membrane B2 is much higher than that of commercial membranes with comparable salt rejection.

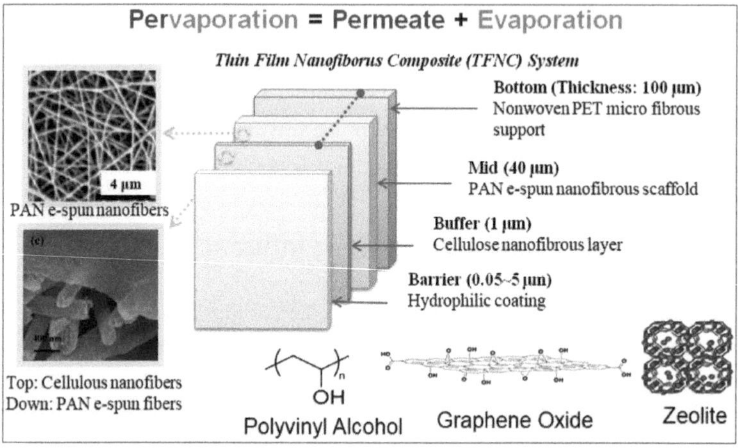

Figure 4.67 Design of novel pervaporation membrane [15, 18].

4.5.5 Pervaporation

It is being mentioned in the current discussion in order to indicate that nanofibrous membranes can also contribute to improving the efficiency of the process. While distillation tends to use more energy and to produce more emissions, **pervaporation** lowers energy waste and is a cleaner process with little emissions (Figure 4.68).

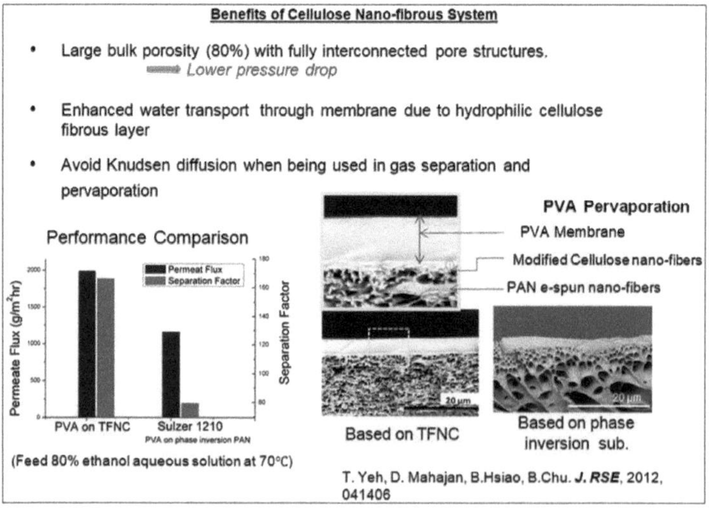

Figure 4.68 Benefits of thin film nanocomposite (TFNC) system [19].

The example is to demonstrate another use of cellulose nanofibers, a product from natural biological systems, mainly trees, and can be useful in the *pervaporation* process [19].

4.5.6 Graphene Oxide Multilayered Membrane Morphology

A sketch of multi-layers of graphene oxide (GO) on top of a cellulose nanofibrous layer has been shown by Sungjin Park and Rodney [20], whereby the authors used SEM, AFM, and TEM to examine the graphene oxide multilayered membrane morphology.

4.5.7 Cellulose Nanofiber Characterization

The characterization of cellulose nanofibers is a more complex issue because of the biological nature of cellulose. In this book, we shall waive the mathematical development and again emphasize the physical picture. As has been mentioned previously, the computational details on SAXS/WAXS data can be found in other books. We provide some preliminary considerations for the reader, so as to get a sense of the challenge and show a couple of results, emphasizing on the bio-complexity of the materials involved in the analysis. For many applications, other means of determining the magnitude of sizes of nanofibers, such as electron microscopy, shall be sufficient.

The procedures for X-ray scattering data start with some assurance that the experimental X-ray scattering data have gone through the standard data correction procedure as shown below. When one can handle the weak X-ray scattering data being comparable to that of pure water, as shown in Figure 4.69, the data should be of very high quality, not easily achievable by inexperienced researchers, dealing with small angle X-ray scattering, such as a beginning graduate student in chemistry. Thus, precautionary measures, for example, by asking advice from more experienced researchers in the field could be a reasonable approach. Please note that the X-ray intensity in Figure 4.69 is on a log scale.

In Figure 4.69, the following is noted:

- The beam center calibration was done using the standard Silver Behenate (d_{001} = 5.84 nm).

- Dead pixels and pixels behind beam stop were blocked by using dark current and mask correction.
- Water scattering together with capillary scattering was subtracted from the suspension scattering.

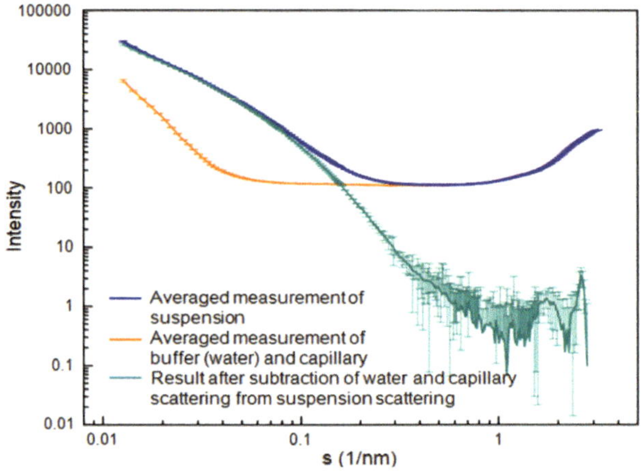

Figure 4.69 SAXS characterization of cellulose nanofibers [21].

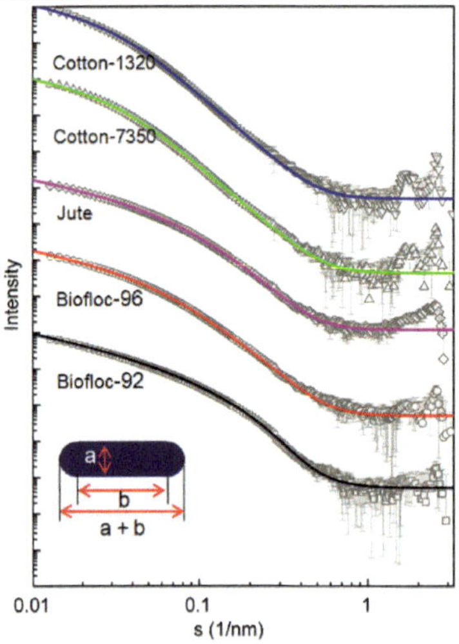

Ribbon model with Gamma distributions of *a* and *b*

Raw Material	Size (Weight Average) (nm)
Biofloc-92	$a_w = 2.2 \pm 0.85$ nm $b_w = 6.1 \pm 5.3$ nm $a_w + b_w = 8.3 \pm 5.4$ nm
Biofloc-96	$a_w = 3.2 \pm 2.2$ nm $b_w = 9.5 \pm 5.0$ nm $a_w + b_w = 12.7 \pm 5.5$ nm
Jute	$a_w = 2.6 \pm 1.6$ nm $b_w = 8.5 \pm 4.0$ nm $a_w + b_w = 11.1 \pm 4.3$ nm
Cotton-7350	$a_w = 7.5 \pm 4.9$ nm $b_w = 12.9 \pm 4.2$ nm $a_w + b_w = 20.4 \pm 6.5$ nm
Cotton-1320	$a_w = 8.6 \pm 5.5$ nm $b_w = 12.9 \pm 4.2$ nm $a_w + b_w = 21.5 \pm 6.9$ nm

Figure 4.70 Analysis of SAXS data with the ribbon model.

In the analysis of SAXS data with the ribbon model (Figure 4.70), only the results of analysis are shown. The experimental results agree very well with the theoretical computations. We can see that different sources of cellulose yield different sizes of well defined nanofibers. Those nanofibers can be used for many applications.

The final example deals with an interesting material (Section 4.5.9). It is natural, highly porous, and inexpensive. The author has personally used it as a bath sponge when he was a boy in Shanghai.

4.5.8 Oil-Absorbent Sponges (Hydrophobic Sponges for Oil Absorption)

Luffa sponges are highly porous and inexpensive, readily available as a *hydrophobic* bath sponge for scrubbing purposes. The fibers are composed of ~60% cellulose, ~30% hemicellulose, and ~10% lignin [22].

One of the ways to modify luffa sponge was to use a 2-step facile method with polyhedral oligomeric silsesquioxane (POSS) as base materials. The hydrophobic behavior can be visualized by looking at the original sponge versus the modified hydrophobic sponge, as shown in Figure 4.71.

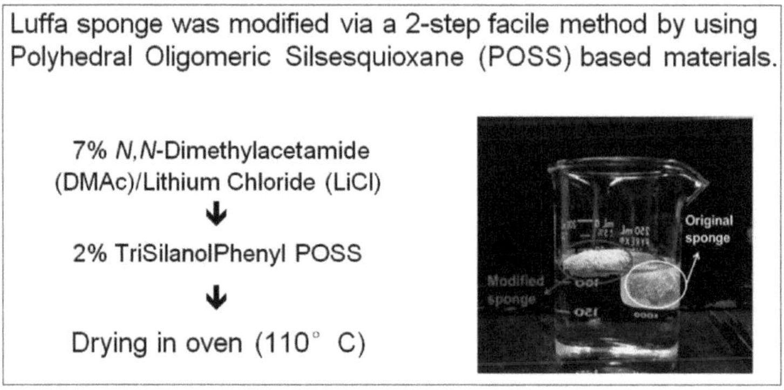

Figure 4.71 Surface modification of luffa sponge.

The super-hydrophobicity of the modified sponge is shown in Figure 4.72.

SEM images and energy-dispersive X-ray spectrum of luffa sponges before and after modification are shown Figure 4.73.

The oil absorption performance of luffa sponge has been carried out by using different types of oil and a couple of organic liquids (DMF and DMSO) (Figure 4.74). The results show that the luffa sponge can absorb 8 to 12 times of its weight for different types of oils.

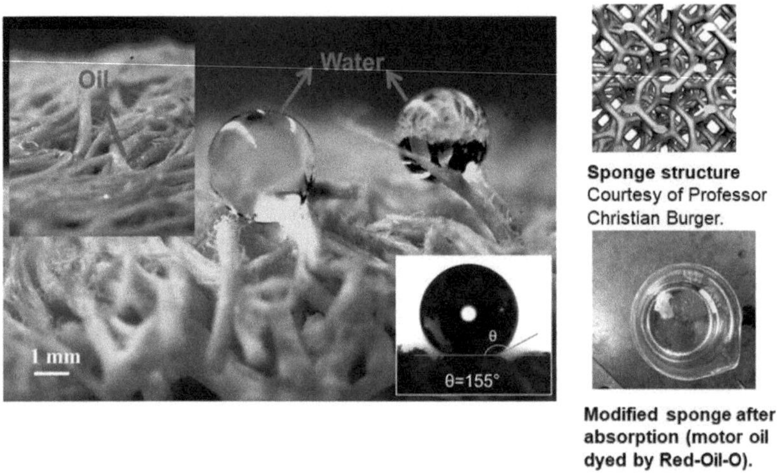

Figure 4.72 Super-hydrophobicity of the modified sponge. (Left) Exposing the modified sponge in oil and in water, (Bottom right) The sponge structure and the hydrophobic sponge after absorption of motor oil with a red dye, (Top right) Schematic presentation of the microstructure of luffa sponge, courtesy of Professor Christian Burger.

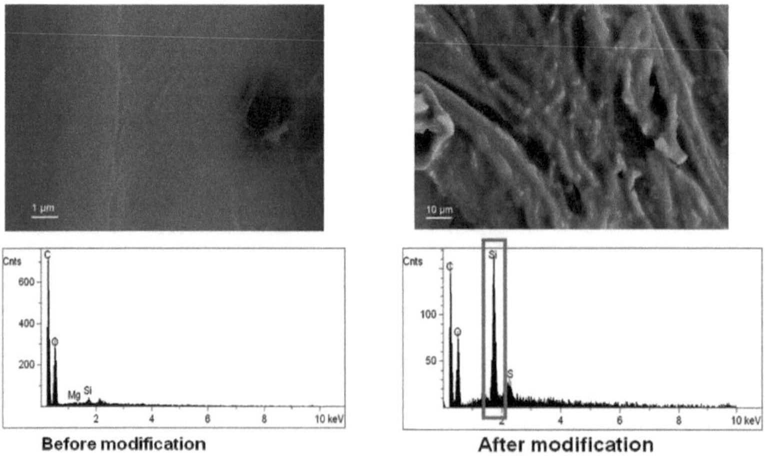

Figure 4.73 Surface morphology of sponge. SEM images and energy-dispersive X-ray spectrum of luffa sponges before (left) and after (right) modification.

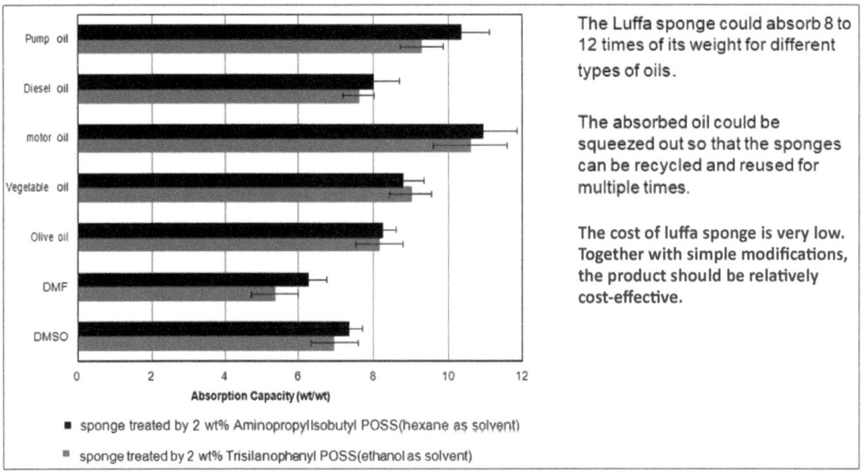

Figure 4.74 Oil absorption performance.

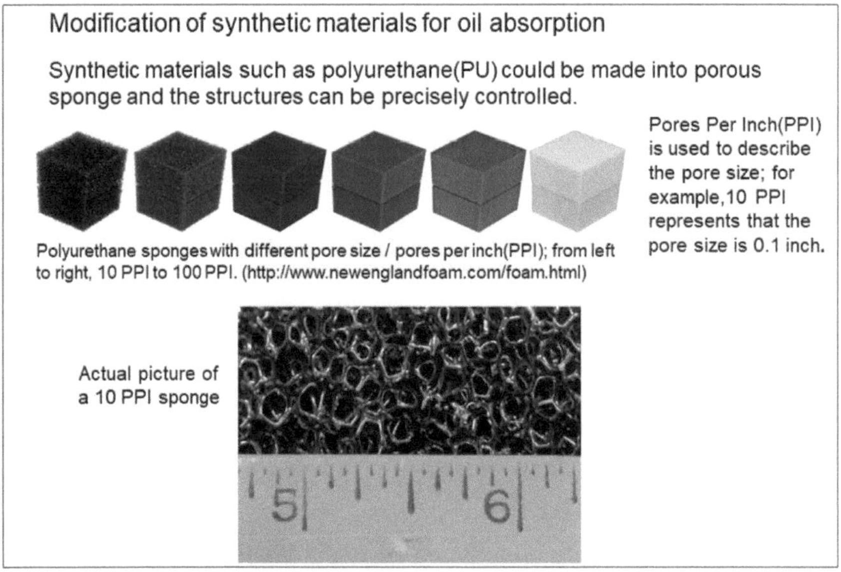

Figure 4.75 Synthetic sponges for oil absorption.

The preliminary test showed the modified polyurethane sponge could take up more than 20 times its weight for motor oil.

The presentation of the two cases on hydrophobic sponges above is to illustrate that in fluid purification processes, there are different approaches and different emphasis on the target(s) involved. Often, there are different needs and sources of impurities, different amounts of impurities and varying local conditions involved. For water purification, the basic environment as related to the water need, sources of water and of impurities, is essentially similar, although the scale may cover a broader range and local conditions vary. Nevertheless, nanofibers seem to have interesting advantages that we should keep its utility in mind.

4.5.9 Summary

1. **Microfiltration membrane**

   Comparing with commercially available counterparts: Flux: 2 ~ 5 times better;

   Retention: 99.9999% against *E. coli*

   99% against MS2

   ~ 3 times higher capacity for toxic metals

   Pressure drop: as low as 0.2 ~ 2 psi

2. **Ultrafiltration (UF) membrane**

   Comparing with commercial available counterparts: Flux: 2 ~ 10 times;
   Retention: > 99.5 % against oily wastewater;
   Anti-fouling: ~ 100% recovery after soap washing.

3. **Nanofiltration (NF) membrane**

   With CN substrate, slot-die coating and co-monomer (BP), NF membrane permeability can be improved by about twice as high as commercial ones (Example, NF270).

4. **Pervaporation**

   TFNC substrate is an ideal substrate for ethanol dehydration by pervaporation due to its high porosity and high hydrophilic properties. PVA/TFNC membrane had about 2× higher permeate flux than commercialized PVA membrane. GO/TFNC membrane was synthesized and applied to ethanol dehydration. GO/TFNC reached 2× higher flux than that of commercialized membranes.

5. **Characterization of SAXS/WAXS cellulose nanofibers**

   SAXS/WAXS of cellulose nanofiber suspensions yielded a polydisperse ribbon model for width and thickness information on fiber cross-section computations.

   XRD data of freeze-dried cellulose nanofibers with modified Debye analysis confirmed ribbon model and provided details about the arrangements of cellulose molecular chains inside the ribbons.

6. **Super-hydrophobic luffa/polymer sponge for cost-effective oil removal**

   SAXS/WAXS of cellulose nanofiber suspensions agrees with a ribbon model for the width and thickness information on the fiber cross-section.

   XRD data of freeze-dried cellulose nanofibers with modified Debye analysis confirmed the ribbon model analysis and provided some details about the arrangements of cellulose molecular chains inside the ribbons.

4.6 Characterization of Cellulose Nanofibers with Applications to Separation Membranes for Water Purification

Our interest in cellulose nanofibers starts with the original primary aim of water purification. However, water is essential to our well-being and has a broader perspective to our society. One relationship is the concept "among water-energy-environment". We can think of water for energy and energy for water. The upper and lower halves of the circle (water for energy and energy for water) implies that

- water is needed to generate energy and energy is needed to deliver clean water,
- water and energy are deeply coupled with each other, and
- water or/and energy may be running short in many regions in the world.

4.6.1 Water-Energy Nexus (water-Energy-Environment)

When we discuss the water purification process, the implications are much broader and are interrelated. The second point is that water purification often covers a very broad range of undesirable spies that we want to remove. With the final "purified" water depending on what we want the "purified" water for. For example, requirements for drinking water and agricultural water are not the same. The amount and type of impurities and the amount of useful water that we need for a specific purpose are subject to cost analysis. Thus, we can see that water purification is a complex process. At the same time, the complexity offers opportunities. It is a very exciting topic, worthy of our attention and effort for the benefit of society.

When we start to consider water purification, we first think of the impurities in water. Thus, an amount of background information on the classification of water impurities suitable for membrane filtration is in order. The classification of membrane filtration covers sizes ranging from water (H_2O) molecules to the yeast cell and the corresponding membrane filtration processes from RO to conventional filtration. The corresponding range of driving force and membrane pore size has provided a very good overall picture on what the magnitude of the problem we are dealing with. In the size exclusion range, we have RO with sizes of <1 nm, nanofiltration (NF) with sizes of 1–10 nm, ultrafiltration (UF) with sizes of 10–100 nm and microfiltration (MF) with sizes of 0.1–50 micrometers. At the same time, smaller pore sizes imply the need for higher applied pressure and lower flux for the fluid to pass through the filter.

In comparison with conventional water filtration membranes (since the seventies), there have been substantial improvements. The use of nanofibers and the introduction of cellulose-type of biomaterials for filtration membranes represent two remarkable changes in recent years. However, the complexity of biological systems again offers new opportunities, especially when the materials can be intermingled or form complexes with synthetic organic/inorganic materials. Thus, more breakthroughs will be anticipated and expected.

We can see how cellulose molecules, nanofibrillated fibers, micro-fibrillated fibers, and wood cell wall, with the size ranging from Angstroms for a cellulose molecule to the nanometer size for nanofibrillated cellulose fibers, micron-sized cellulose fibers and then wood cells, end up forming a tree.

The information on cellulose suggests to us the diversity of cellulose-type biomaterials which can be useful as basis materials, in addition to the electro-spun/blowing nanofibers, primarily based on organic/inorganic chemicals. Biological systems can be more complex and, at times, less well-defined. However, their complexity can be used to our advantage. Thus, sources from plants also offer opportunities.

The information on cellulose suggests to us the diversity of cellulose-type biomaterials which can be useful as basis materials, in addition to the electro-spun/blowing nanofibers, primarily based on organic/inorganic chemicals. Biological systems can be more complex and, at times, less well-defined. However, their complexity can also be used to our advantage. Thus, sources from plants offer additional opportunities.

In the preparation of ultrafine cellulose nanofibers, cellulose microfibers from wood pulp go through a series of chemical steps, as shown in Figure 4.76, with wood pulp starting at fairly large sizes of ~40 μm in fiber diameter down to cellulose nanofibers with fiber diameters in the nm-size scale.

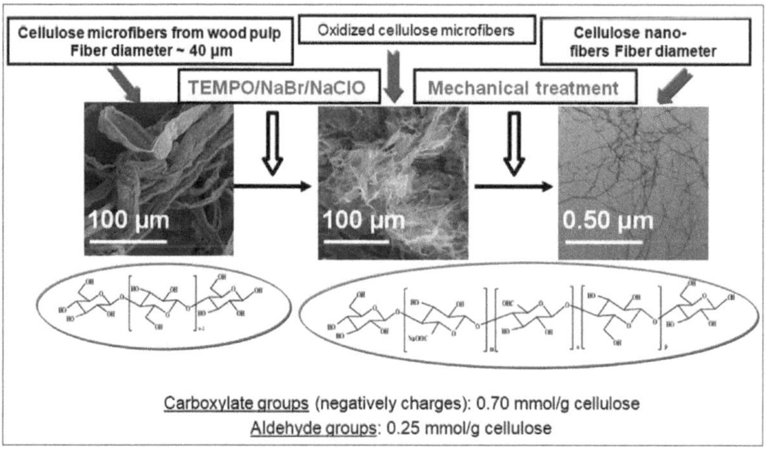

Figure 4.76 Preparation of ultrafine cellulose nanofibers.

We have prepared a series of cellulose nanofibers from a range of cellulose raw materials sources. The preparation method starts with the source in the pulp form (Figure 4.77). Then, cellulose nanofibers were further purified by dialysis to retain a fixed range of molecular weight. As having been discussed earlier, the details in the preparation and characterization are being bypassed. Those interested in looking into more details further are being referred to references [15, 21].

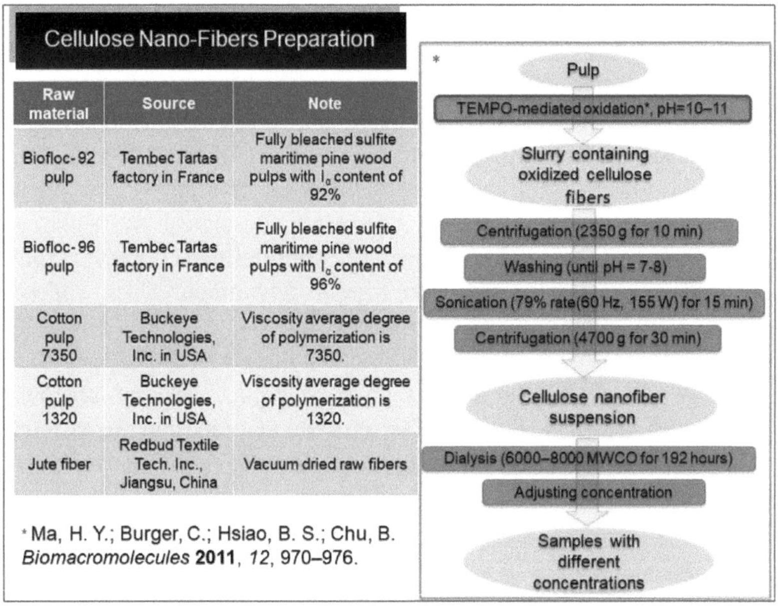

Figure 4.77 Cellulose nanofiber preparation [15].

In addition to the examples of source materials for cellulose nanofibers from a range of cellulose raw materials sources, consisting of biofloc 92 and 94 pulp, cotton linter 7350 and 1320, and jute fiber, as shown in Section 4.5, it may be interesting to consider falling leaves as a possible source material. A substantial amount of cellulose components are present in many plants. Thus, the key decision continues to depend on the specific purpose and the economic consideration.

For more SAXS characterizations, see references [7, 21]

Figure 4.78 shows the model fitting results of using the ribbon model in comparison with experimental measurements. The left panel shows a comparison of different fiber (cotton, spruce, jute and bamboo from SAXS measurements in comparison with the ribbon model. The right panel shows electron micrographs of cotton, bamboo and spruce-I & -II fibers, with substantial differences in cross sections.

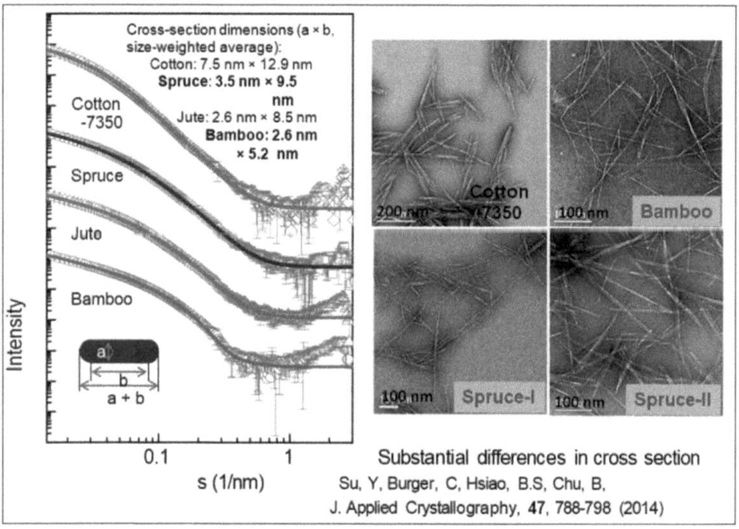

Figure 4.78 Model fitting of ribbon model to experimental data [21].

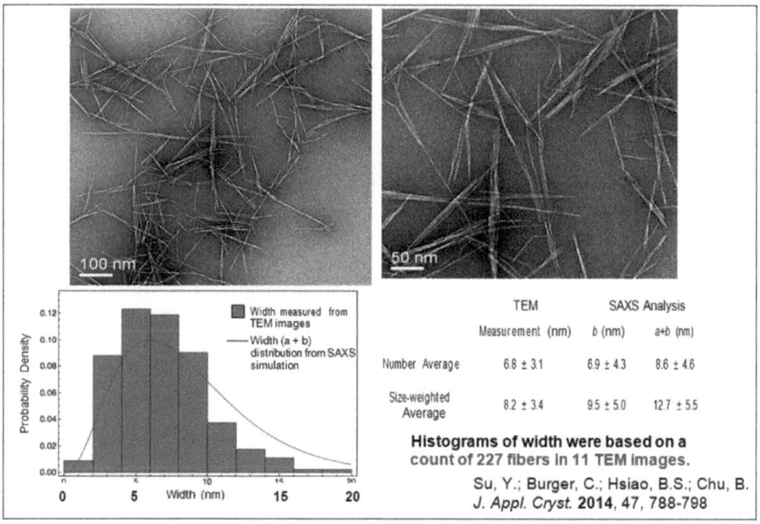

Figure 4.79 TEM characterization of cellulose nanofibers [21].

Figure 4.80 shows AFM measurements of nanostrips.

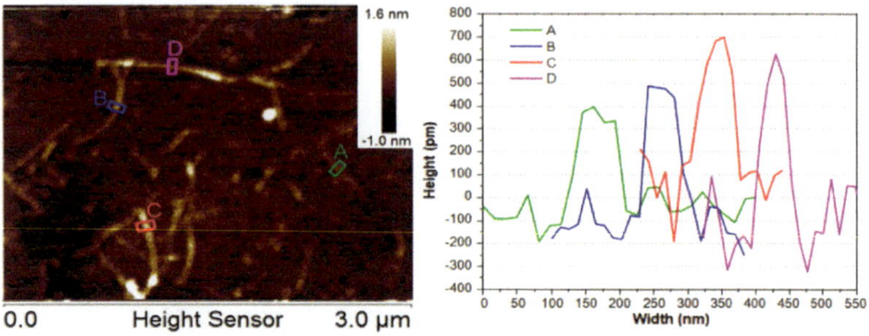

Figure 4.80 AFM measurements of nanostrips. For most fibers in the image, the thicknesses were less than 1 nm.

It is worthwhile to keep in mind the use of cellulose nanofibers for filtration applications. Apart from their existence as natural materials and plant nanofibers, such as cellulose, they can be formulated without using the electro-spinning/-blowing processes. An example is our recent development on their application to membrane distillation.

4.6.2 Technology Advantage Summary

High flow rate: 2–10 times of competitive nanomaterials and structures

- Tensile strength, Porosity, Pore size
- Superior surface modification capability
- Unique process and equipment
- Durable, novel and low cost raw materials
- Commercial grade PET as supporting structure
- Consistent electro-spinning layer
- Barrier layer with ultrathin nanofibers

Platform Technology

- Broad application with the same platform technology
- High flow rate/Low pressure drop save energy and reduce operation cost
- Enable new applications, such as portable or off-shore smaller systems

Overview

It may be of interest for the reader to view a summary of this book, primarily relating to Chapters 2–4:

Chu has presented Chapter 2 dealing with the development of themes related to laser light and synchrotron X-Ray scattering, Chapter 3 on the evolution of electro-spinning technology and Chapter 4 on its potential applications of nanofibers, with a short Chapter 5 on a proposed application to biomedical themes of importance.

The chapters are correlated, i.e., the themes are related to one another for their utility and existence in this book. The reader may wish to know that in describing the capability of laser light and synchrotron X-ray scattering, Chu has relied on experimental evidence from the able Chu team at SBU to elucidate each capability of laser light/synchrotron X-ray scattering by carrying out individual experiments, independent of results from the literature. Thus, we know that laser light scattering can be combined with synchrotron X-ray scattering measurements (Figure 2.38), and laser light scattering can be carried out by using a cylindrical cell of 1–2 mm diameter (Figures 2.49–2.51).

In Chapter 4, two long-standing problems on (1) coil-to-globule translation and (2) behavior of plasmid DNA were selected to demonstrate the capabilities of a combined light-X-ray technique, as being partially illustrated in Figure 4.21, on the enormous volume change and the dramatic conformation from a random coil to a molten globule. It is important to understand that these are examples being used to demonstrate the capability of the technology, which are based on fundamentals.

References

1. Z. M. Huang, Y. Z. Zhang, M. Kotaki, S. Ramakrishna, *Composites Science & Technology* 63, 2223–2253 (2003). This is also a noteworthy review on polymer nano-fibers by electro-spinning and their applications in nano-composites.
2. C. Burger, B. S, Hsiao, and B. Chu, *Annual Review of Materials Research*, **36**, 333–368 (2006).
3. C. Burger, S. Q. Zhou, and B. Chu, in: *Handbook of Polyelectrolytes & Their Applications*; American Scientific Publishers, chapter 7, pp. 125–141 (2002).

4. K. Luu, K. Kim, B. S. Hsiao, B. Chu, M. Hadjiargyrou, *J. Control Release,* **89**, 341–353 (2003).
5. E. Stanley, *Physics Today,* **56**, 40–46 (2003).
6. H. Y. Ma, C. Burger, B. S. Hsiao, B. Chu, *Journal of Materials Chemistry,* **21**(21), 7507–7510 (2011).
7. Ying SU, PhD thesis, Structure Characterization of Cellulose Nanofibers/Microfibrils, Stony Brook University, 12/2014, including several figures on cellulose structures.
8. H. Y. Ma, C. Burger, B. S. Hsiao, B. Chu, *Biomacromolecules,* **12**, 970–976 (2011).
9. A. Sato, R. Wang, H. Y. Ma, B. S. Hsiao, B. Chu, *Journal of Electron Microscopy,* **60**, 201–209 (2011).
10. H. Y. Ma, B. S. Hsiao, B. Chu, *ACS Macro Letters*, 2012, **1**, 213–216.
11. H. Ma, C. Burger, B. S. Hsiao, B. Chu. *ACS Macro Letters,* **1**, 723–726 (2012).
12. B. Chu, B. S. Hsiao, *Journal of Polymer Science Part B: Polymer Physics,* **47**, 2431–2435 (2009).
13. H. Y. Ma, C. Burger, B. S. Hsiao, B. Chu, *Bio-macromolecules,* **13**, 180–186 (2012).
14. A. Sato, R. Wang, H. Y. Ma, B. S. Hsiao, B. Chu, B. *Journal of Electron Microscopy,* **60**, 201–209 (2011).
15. H. Y. Ma, C. Burger, B.S. Hsiao, B. Chu, *Biomacromolecules,* **12,** 970–976 (2011).
16. J. A. Otero, O. Mazarrasa, J. Villasante, et al. *Journal of Membrane Science,* **309**, 17–27 (2008).
17. A. Tiraferri, M. Elimelech, *Journal of Membrane Science,* **389**, 499–508 (2012).
18. B. Chu, B. S. Hsiao, H. Y. Ma, WO 2010/042647; PCT/US09/059884.
19. T. Yeh, D. Mahajan, B. Hsiao, B. Chu, *Journal of Renewable and Sustainable Energy,* **4**, 041406 (2012).
20. S. Park, R. S. Ruoff, *Nano Letters,* **2**(3), 572–578 (2007).
21. Y. Su, C. Burger, B. S. Hsiao, B. Chu, *Journal of Applied Crystallography,* **47**, 788–798 (2014).
22. B. Chu, B. Hsiao, "Nanostructured Fibrous Membranes for Membrane Distillation," via Stony Brook University Research Foundation, issued on March 17, 2023, US Patent 11,596,886.

Chapter 5

Aims and Looking Forward

This book has two aims as indicated by its title:

1. A combination of laser light and synchrotron X-ray scattering can provide us a useful tool to investigate complex dynamic problems. An example may deal with kinetic studies on the aggregation of species, such as early stages of amyloid fibrils in Alzheimer's, and
2. The development of ultrafine fibers by using either electro-spinning/blowing technology or from biological systems, such as cellulose from trees, for water (air) purification.

Toward the end of Chapter 2, we have raised the topic on how laser light and synchrotron X-ray scattering can be combined to investigate *amyloid fibril formation* in the very early stages with input information from other measurements on its aggregation behavior. It should be noted that tests of known samples, based on the angular and wave length ranges of lasers and synchrotron X-rays, showed that a wide range of sizes and time scales are within range of such a setup. It should also be noted that while laser light and synchrotron X-ray scattering measurements are powerful tools, the capability of other physical techniques, such as NMR, microscopy with different forms of probing radiation, are useful when combined together with laser light and synchrotron X-ray scattering studies for complex systems of interest.

Laser Light and X-Ray Scattering: With Evolution of Electrospinning Technology in Biomedical Applications and Water Purification
Benjamin Chu

ISBN 978-981-4968-88-1 (Hardcover), 978-1-003-54074-8 (eBook)
www.jennystanford.com

The pathways of aggregation on the structure, formation and inhibition of amyloid fibrils were presented by Schnabel [1], a science writer, more than 10 years ago. Individual amyloid-β peptides can assemble in at least two ways:

1. One pathway leads to insoluble, plaque-forming fibrils
2. The other way leads to soluble oligomers.

The insoluble oligomers are suspected to be the main toxic species in Alzheimer's disease.

The reader is advised to read the *Nature* article by Schnabel [1], who described the discussions, speculation, and concerns of amyloid fibrils as related to Alzheimer's disease.

The combination of light and X-ray probes (as discussed in Chapter 2) may provide us with another powerful tool that can add new insight for a better understanding of such diseases. An example is to consider the works of Robert Tycko and his coworkers. By using solid state NMR and different forms of microscopy, they have studied beta-amyloid fibrils in Alzheimer's disease. Tycko and his team have done extensive studies on this topic. Example publications are given in references [2–4]. The results suggest how a simulated set of experiments may be carried out by using other known data from studies of structural formation of amyloid fibrils. The information can then be converted to the scattering behavior of such a system so as to examine whether we can learn dynamic processes as related to amyloid fibril formations, as illustrated in Figure 5.1.

The development of a combined laser light and X-ray scattering instrument, as has been described and tested in Chapter 2, is aimed at kinetic studies on "aggregation" phenomena whereby a limited number of multiple species may undergo changes, such as size, shape, or/and aggregations, as proposed by Tycko.

The reader may be interested to consider the example being presented in Chapter 5 as two parts: One part deals with the *subject of interest*, which, in the present case, is concerned with pathways of oligomer and filament formation for a specific system. In the present example, the subject matter is described by Tycko and his team. The other part deals with laser light/synchrotron X-ray scattering, as presented in Chapter 2, which can be used for a much broader range of applications, not only for the specific application example, as presented in Chapter 5, but also for other biological and medical problems of interest.

It is clear that although laser light and synchrotron X-ray scattering measurements (Figure 5.1) are powerful tools, the range of other physical techniques, such as NMR, microscopy with different forms of probing radiation, are useful when combined together with laser light and synchrotron X-ray scattering measurements in more complex studies.

The development of a combined laser light and X-ray scattering instrument, as has been described and tested in Chapter 2 is aimed at kinetic studies on 'aggregation' phenomena whereby a limited number of multiple species may undergo changes, such as size, shape, or/and aggregations. It is clear that although laser light and synchrotron X-ray scattering measurements are powerful tools, the range of other physical techniques, such as NMR, microscopy with different forms of probing radiation, are useful when combined together with laser light and synchrotron X-ray scattering measurements in more complex studies.

A cartoon on pathways of oligomer and filament formation is being presented to show how a simulated set of experiments may be carried out by using other known data from studies of structural formation of amyloid fibrils. The information can then be converted to scattering behavior of such a system so as to examine whether we can learn dynamic processes as related to amyloid fibril formations.

Figure 5.1 shows how laser light and synchrotron X-ray scattering measurements, in combination with other physical techniques can be carried out to learn more on the initial aggregation behavior of a few species in the initial formation of the Alzheimer disease. In Figure 5.1, a schematic presentation of the instrument configuration for both laser light scattering and synchrotron SAXS/WAXS and date analysis is shown in the bottom panel. It should also be noted that the laser light scattering set up uses twin lasers with two twin APDs for the proposed studies. The different simulations based on monomers, dimers and trimmers are shown as red balls in the middle top panel. The measured dynamic light scattering (DLS) results at two different scattering angles (here denoted as 30° and 150°) are shown on the left panels, while composite X-ray scattering curves are shown on the right panel.

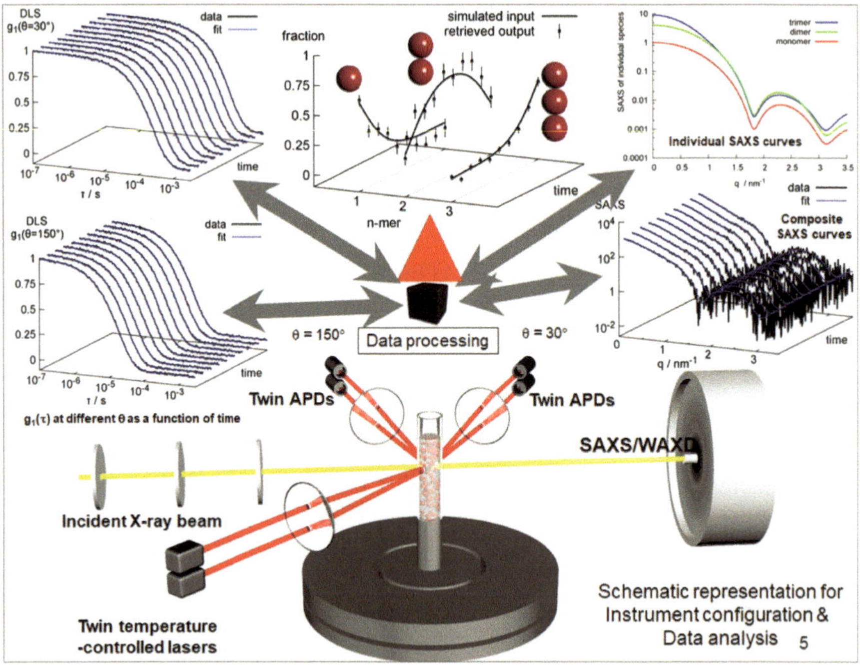

Figure 5.1 Laser light and synchrotron X-ray scattering measurements.

Looking Forward

The use of cellulose-based materials from plants offers many alternative pathways for water (air) purifications. Its strength lies in the possibility for surface modifications and lower costs for a specific application. On the other hand, for biomedical applications where high quality may be of primary importance and cost becomes a secondary consideration, electro-spun nano-fibers will likely become a useful pathway.

The reader may be interested to know that the most important intention of the author is to try to introduce the two topics to beginning graduate students in terms of basic physics, leaving out the mathematical details, so that the user of such instruments or materials has a better idea on some of the basic principles.

The future is challenging with COVID viruses, high population density, climate change, and uncertain politics. Yet, science and engineering remain exciting and provide us with new opportunities to contribute to our society and to help us live a better life. Aside from the digital world coming our way, the topics of this (introduction) book also offer us new approaches from a different angle to be happy and healthy persons who can also help others having a better life.

References

1. J. Schnabel, *Nature*, **475**, S12–S14 (2011) https://doi.org/10.1038/-475S12a.
2. A. K. Paravastu, A. T. Petkova, R. Tycko, *Biophysical Journal*, **90**(12), 4618–4629 (2006).
3. J.-X. Lu, W. Qiang, W.-M. Yau, C. D. Schwieters, S. C. Meredith, R. Tycko, *Cell*, **154**(6), 1257–1268 (2013).
4. W. Qiang, W.-M. Yau, J.-X. Lu, J. Collinge, R. Tycko, *Nature*, **541**(7636), 217–221 (2017).

Acknowledgements

Aside from specific contributions described on page 22, first paragraph, BC wishes to formally thank many team members for their valuable efforts in completing this book: More specifically, Walther Tscharnuter, Bruce Weiner, and Dufei Fang.

The two primary topics of this book were developed with efforts and thoughtful contributions by many members of my research team at Stony Brook over the years, especially those involved in studies related to cellulose structures, including Ying Su, Zhe Wang, Yimin Mao, and Professor Hongyang Ma. In particular, I would like to acknowledge my appreciation to Professor Dufei Fang for his work on laser light and synchrotron X-Ray scattering instrumentation, Professor Christian Burger for his insight on water channels in cellulose materials, Professor Chi Wu for his thoughtful clarification on the assembly of polymeric materials, and Professor Tianbo Liu for his lead contribution in setting the landscape of the two primary reviews.

I am thankful to Brandon Huynh, Vi computer software adviser, for his patience, knowledge, and understanding in computer technology, resolving many presentation details in formulating the book to its final form.

Index